14th International Colloquium on Paratuberculosis (ICP 2018)

Riviera Maya, Mexico
4 – 8 June 2018

ISBN: 978-1-5108-7302-5

Printed from e-media with permission by:

Curran Associates, Inc.
57 Morehouse Lane
Red Hook, NY 12571

Some format issues inherent in the e-media version may also appear in this print version.

Printed by Curran Associates, Inc. (2019)

For permission requests, please contact International Association for Paratuberculosis at the address below.

International Association for Paratuberculosis
2015 Linden Drive
Madison, Wisconsin 53706-1102
USA

Phone: (608) 262-8457
Fax: (608) 265-6463

www.paratuberculosis.org

- **Additional copies of this publication are available from:**

Curran Associates, Inc.
57 Morehouse Lane
Red Hook, NY 12571 USA
Phone: 845-758-0400
Fax: 845-758-2633
Email: curran@proceedings.com
Web: www.proceedings.com

TABLE OF CONTENTS

Welcome message

Dear members of the IAP,

I'm pleased to inform you that, as approved by the IAP, in 2018 Mexico will host the 14th Colloquium of the International Association for Paratuberculosis in the Riviera Maya, Quintana Roo.

Cancun-Riviera Maya is one of the most dynamic cultural capitals, rich in history and alive with excitement and friendliness people, where visitors each year come from all parts of the world, making this the best showcase for all that Mexico has to offer.

Also, in a veterinary perspective, the Faculty of Veterinary Medicine and Animal Science of the National Autonomous University of Mexico –which is rated fourth in the Latin America rankings– has the academic expertise to compete with the higher standards that have been set by IAP through the time.

The time proposed for the Colloquium is the beginning of June, which is the beginning of the summer in Mexico when the days are longer and the weather is wonderful.

The International Colloquium on Paratuberculosis, periodically held under the auspices of the International Association for Paratuberculosis, has become a special meeting for the IAP members and, more importantly, for leading researchers, livestock industry representatives, veterinarians and public health authorities with an interest in this disease and its related issues, scientific sessions will focus on the biology of MAP, pathogenesis and immunology, diagnostics, national control programs, epidemiology and control strategies, concluding with and update on public health and food safety aspects.

The National Autonomous University of Mexico (Universidad Nacional Autónoma de México) and the Sociedad Mexicana de Paratuberculosis y otras micobacteriosis A. C., are honored to receive the trust of the IAP and lead the organizational efforts to make the 14th colloquium an event of scientific interest and social enjoyment.
I am looking forward to greeting you in Riviera Maya, in June 2018.

Dr. Gilberto Chávez Gris
Congress Chairman of the 14th ICP

Organizing Committee

Local committee

Gilberto Chávez Gris (Chairman)
President of the Mexican Society of Paratuberculosis and other Mycobacteriosis
Veterinary Medicine and Zootechnics
Universidad Nacional Autónoma de México (UNAM)
gris@unam.mx

Dr. Jorge Tórtora Pérez
Professor of the Department of Higher Education of Cuautitlán of
Universidad Nacional Autónoma de México (UNAM)

Carolina Segundo Zaragoza
Veterinary Medicine and Zootechnics
Universidad Nacional Autónoma de México (UNAM)

Victoria Castrellón Ahumada
Universidad Autónoma de Zacatecas "Francisco García Salinas", Unidad Académica de Medicina Veterinaria y Zootecnia.

Karina Cirone
Department of Animal Production
Instituto Nacional de Tecnología Agropecuaria (Argentina)

Claudia Morsella
INTA-Balcarce
Department of Animal Production
National Institute of Agropecuarian Technology (Argentina)

Edith Maldonado Castro
Treasurer of the Mexican Society of Paratuberculosis and other Mycobacteriosis
Veterinary Medicine and Zootechnics
Universidad Nacional Autónoma de México (UNAM)
macae09@hotmail.com

Ángel Pulido Albores.
Member of the Mexican Society of Paratuberculosis and other Mycobacteriosis
Veterinary Medicine and Zootechnics
Department of
Universidad Nacional Autónoma de México (UNAM)

Gabriel Ernesto Pallás
Member of the Mexican Society of Paratuberculosis and other Mycobacteriosis
Agricultural Sciences center of
Universidad Autónoma de Aguascalientes

Erika Torres Ramos
Private Excercise

Fernando Policchi
INTA-Balcarce
Private Excercise
Universidad Nacional de Mar del Plata (Argentina)

Arnulfo Villanueva Castillo
Benemérita Universidad Autónoma de Puebla

Ruby Sandy Moreno Mejía
Secretary of the Sociedad Mexicana de Paratuberculosis y otras Micobacteriosis A.C.
Benemérita Universidad Autónoma de Puebla

Gilberto Ballesteros Rodea
Member of the Sociedad Mexicana de Paratuberculosis y otras Micobacteriosis A.C.
Agronomy and Veterinary Department of Universidad Autónoma de San Luis Potosí

Board of directors

IAP Board

Officers of the association

Scientific Committee

Session 1
Diagnostic, Immunology and Host Response

Edith Maldonado Castro (Mexico) *edihtmc@unam.mx*
Silvia Mundo (Argentina) *s_mundo2004@yahoo.com.ar*
Eiichi Momotani (Japan) *eiichimomotani@gmail.com*
Ray Sweeney (USA) *rsweeney@vet.upenn.edu*
Valentín Pérez (Spain) *vperp@unileon.es*
Gregers Jurgensen (Denmark) *grju@vet.dtu.dk*

Session 2
Map Control Programs and Epidemiology

Fernando Paolicchi (Argentina)
paolicchi.fernando@inta.gob.ar
Joseba Garrido (Spain) *jgarrido@neiker.eus*
Christine Fourichon (France) *christine.fourichonl@oniris-nantes.fr*
Nicola Pozzato (Italy) *npozzato@izsvenezie.it*
Miguel Salgado (Chile) *miguelsalgado@uach.cl*
Cristobal Verdugo (Chile) *cristobal.verdugo@uach.cl*

Session 3
Pathogenomics and Map Biology

Victoria E. Castrellón Ahumada (Mexico)
vecka_20@hotmail.com
Karen Stevenson (UK) *karen.stevenson@moredun.ac.uk*
John Bannantine (USA) *john.bannantine@ars.usda.gov*
Adel Talaat (USA) *adel.talaat@wisc.edu*

Session 4
Public Health and Map in the Environment

Karina Cirone (Argentina) *cirone.karina@inta.gob.ar*
Norma Arrigoni (Italy) *norma.arrigoni@izsler.it*
Leonardo Sechi (Italy) *sechila@uniss.it*
Tim J. Bull (UK) *tim.bull@sgul.ac.uk*
Alessia Galiero (Italy) *alessiagaliero@gmail.com*

14 ICP AIP Award

Emeritus
Douwe Bakker (Netherlands)
Murray E. Hines (USA)

Helping Hands Awards
José Miguel Hernández Agudelo (Colombia)
Ganesh Gangarami Sonawane (India)
Mukta Jain (India)
Amit Kumar Singh (India)
Isis Espeschit (Brazil)

Richard S Merkal Awards:
Lucy Luo (Canada)
Caroline Corbet (Canada)

General information

Conference venue

International Convention Center - Hard Rock Hotel Riviera Maya
Chetumal, Carr. Cancún - Tulum, Km 72
Riviera Maya, Quintana.Roo.
ZIP Code 77710.

All the scientific sessions, the IAP General meeting, the IAP Board of directors, the Iberoamerican Forum on Paratuberculsis and the Paratuberculosis Forum will be held at the conference venue.

Poster presentation

In preparing your poster presentation please remember:

- Poster Exhibition session during the conference, which will create an opportunity for networking.
- During the session, presenters are expected to stand by their poster/exhibition in order to discuss their research/design project with the viewers.
- The poster presentations will be on continuous display throughout the conference.
- Poster authors are responsible for printing and bringing the poster to the conference
- Poster set-up: Each poster presenter is responsible for putting his/her poster up on a dedicated board.
- Poster size: The poster should not exceed A0 format (841mm x 1189 mm / 33" x 47") in portrait (vertical) layout
- Poster size: No pins (double-sided scotch will be provided by the organizer)

Authors are encouraged to:
a) Include a photo of themselves on the poster so that they can be easily identified by the poster viewers.
b) Print several copies of their posters in A4 or letter format and hang these up in an envelope on their designated poster board. This will make communication and future collaborations easier and more visible.

Oral presentation

In preparing your oral presentation please remember:
You have 10 minutes for presentation and 5 minutes for questions and comments.
Your presentation must be in format Power Point.
You must be deliver it the day of your presentation in advance of 45 minutes pursuant to the final program, in order to prepare your communication without any mishaps.

Certificate of attendance

A Certificate of Attendance will be provided to each delegate at the conference.

Name Badges

For security purposes, and to facilitate collaboration, delegates, speakers, sponsors, and exhibitors are asked to wear their name badges to all sessions.
If you misplace your name badge, please enquire at the Registration Desk to organize a replacement.

Registration desk

The Registration Desk is located at the main entrance. It will be open at the following times:
Monday, 4 June 2018 11:30
Tuesday, 5 June 2018
Wednesday, 6 June 2018

Thursday, 7 June 2018
Friday, 8 June 2018

Proceedings

Proceedings are currently available on the Colloquium website (www.14thicp.mx).

Internet

There is free wifi in the conference center

Mobile phones

Please switch off your mobile phones during sessions.

App for mobile phone for the 14th International Colloquium of Paratuberculosis

An application for mobile phone for the Colloquium is available, you can follow the download instructions in this link: https://m.youtube.com/watch?v=xGoRUJY18Hs

Currency and money exchange

The monetary unit in Mexico is the Peso mexicano. Currency can be exchanged at the airport, at banks or from cash dispensers. All major credit cards such as MasterCard, Visa and American Express are accepted in the instalations of the Hard Rock Café Hotel.

Hotel

Hard Rock Riviera Maya hotel has 1,264 hotel rooms within 5 minutes walking distance from Internatonal Convention Center Riviera Maya.

Transportation

For getting around you can choose from renting a car, using taxis or using public transportation. The main car rental companies can be found in Riviera Maya and their services can be hired in their own agencies, the airport and some hotels. Taxis offer impeccable service and are available in all hotel lobbies; they do not use meters, but have set rates that are on display in all hotel lobbies. Taxis are available 24 hours a day.

Climate

The average annual temperature of the Riviera Maya is 25.5 degrees Celsius [78 degrees Fahrenheit], with fluctuations of 5 to 7 degrees. In June, July, and August visitors can expect hot sunny summer weather with occasional rainfall. This is also when the sea is the calmest.

Social activities

Welcome reception

Monday, June 4th, from 19:00 to 20:00 h

Afernoon free

Wednesday, June 6th, from 15:00

Gala Dinnner

Thursday, June 7th, from 19:00 to 22:00 h

Things to do in Riviera Maya

Riviera Maya is the new face of Mexico. A destination renewed from its white sand beaches. The Mexican Caribbean is a safe destination. Your hosts are friendly and warm people. The approximately 9 million visitors (congress attendees and tourists) are proof that visiting our beautiful tourist destination is the right choice.

Cenotes: One not-to-miss experience in the Riviera Maya is sweimming in a cenote, a natural cave filled with water. Considered sacred places by the ancient Maya (cenote means "sacred well"), they are found throughout the region. Some of the top spots include Cenote Azul (known for its fresh turquoise water), Gran Cenote (go early to avoid the crowds), and Dos Ojos (where you can snorkel amid stalactites and stalagmites).

Cenote azul

Arecheological ruins: The most notable relics are the monumental constructions and impressive pyramids they built in their religious centres. Chichén-Itzá, which in the Mayan tongue means "on the edge of the well of the Itzáes" is the most important archaeological site in the Yucatan Peninsula and one of the most representative relics of the Mayan culture. In 1988 the United Nations Educational, Scientifific and Cultural Organization (UNESCO) named this archaeological site a World Heritage site and 9 years later, in 2007.

Tulum

Colloquium partners and sponsors

PLATINUM

CNG
Confederación Nacional de Organizaciones Ganaderas

GOLD

IDEXX LABORATORIES

CZV
CZ VETERINARIA

SILVER

IDvet
Innovative Diagnostics

Biolife

STAND

QIAGEN

INSTITUTIONAL PARTNERS

Coordinación General de Ganadería.
SAGARPA

OTHER SPONSORS

Municipio San Juan del Río, Querétaro

GANADERIA RIVERA DE LAJAS

Ganadería Santa Cruz de Xarama

General program

Monday, june 4th	
9:00-12:40	IDF Paratuberculosis Forum (by invitation)
11:30	Registration Opens
12:50-14:50	Foro Iberoamericano (by invitation)
15:00-19:00	IAP Board Directors
19:00-20:00	Welcome Reception
Tuesday, June 5th	
8:00- 8:45	Open Ceremony
Section. Diagnostic, Immunology and Host Response	
8:45-9:30	Lecture. Dr. Claus D Buergelt (U.S.A.): *"Aspects of Pathology and Pathogenesis of Domestic Ruminant Paratuberculosis"* .
9:30-9:45	Coffee break
9:45-12:45	Oral sessions
13:00-14:45	Lunch time
14:45-16:00	Oral sessions
16:00-16:35	Coffee break
16:35-19:00	Poster Sessions
20:00	Dinner
Wednesday, June 6th	
Section. Map Control Programs and Epidemiology	
8:00- 8:45	Lecture. Dr. Richard Whittington (Australia): *"International Control Programs for Paratuberculosis who, how and why?"*
8:45-10:45	Oral Sessions

10:45-11:00	Coffee break
11:00-12:15	Oral Sessions
12:15-14:00	Poster Sessions
14:00-15:00	Lunch time
15:00	Afternoon Free
Thursday, June 7th	
8:00-10:10	Foro Iberoamericano (By invitation)
Section. Pathogenomics and Map Biology.	
10:30-11:15	Lecture. Dr Jeroen De Buck (Canada): *"Recent developments in pathogenomics to understand Map (Mycobacterium avium subsp. paratuberculosis) pathogenesis and Diagnose Johne's disease"*.
11:15-11:30	Coffee break
11:30-13:00	Oral sessions
13:00-14:30	Lunch time
14:30-15:15	Lecture. *Ramón A. Juste Jordan* (Spain): "*Chronic regional intestinal inflammatory disease: A walk through the species*".
15:15-15:30	Coffe break
15:30-16:10	"Richard S. Merkal Award"
16:10-17:00	Poster sessions
17:00-18:00	General Meeting (Only delegates)
18:00-18:15	Group Photography
19:00–22:00	Gala Dinner
Friday, June 8th	
Section. Public Health and Map Environment.	
8:00- 8:45	Lecture. Dr. Rodrick Chiodini (USA): *"Crohn's Disease and Johne's Disease: Why everyone doesn't believe you"*

8:45-10:00	Oral Sessions
10:00-10:15	Coffee break
10:15-11:15	Poster sessions
11:15-12:15	Synopsis of the Colloquium, next 15th ICP.

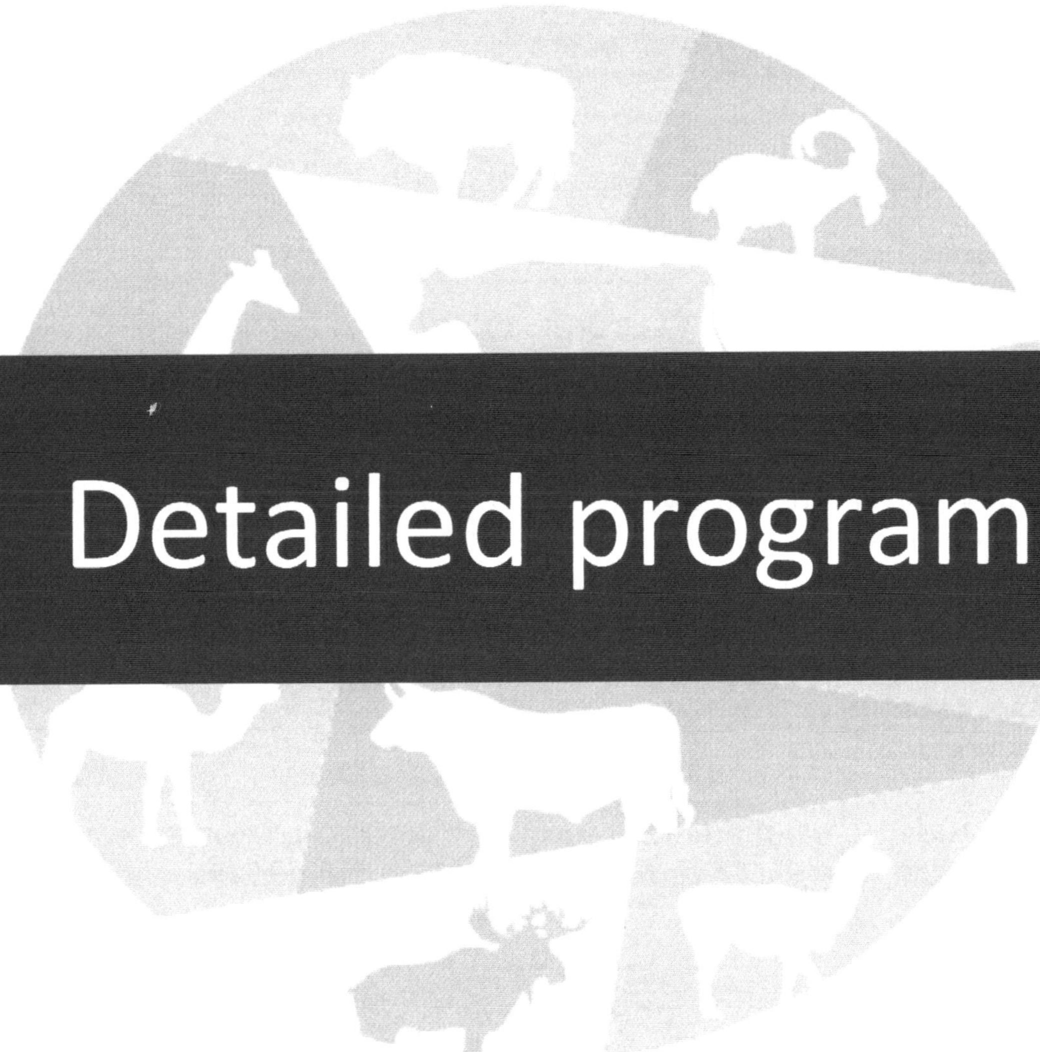

Detailed program

TUESDAY JUNE 5th

Session 1. Diagnostic, Immunology and Host Response

8:45-9:30 **Lecture. Dr. Claus D Buergelt (U.S.A.): "ASPECTS OF PATHOLOGY AND PATHOGENESIS OF DOMESTIC RUMINANT PARATUBERCULOSIS".** 27

9:30-9:45 Coffee break

Moderators: Victoria E. Castrellón Ahumada & Miguel Salgado

1st Oral Session. Moderators: Marta Alonso-Hearn & Frank Griffin (9:30-12:45)

9:45-10:00 **1.1 LIMITED ADAPTIVE IMMUNE RESPONSES OF GOAT KIDS TO MAP BACTERIN VACCINATION WHEN RAISED IN A MYCOBACTERIUM AVIUM SSP PARATUBERCULOSIS (MAP) FREE ENVIRONMENT.** Ad Koets[1], Lars Ravesloot[1], Robin Ruuls[1], Annemieke Dinkla[1], Karianne Lievaart-Peterson[2]. 30
[1] Wageningen Bioveterinary Research, Lelystad, The Netherlands
[2] GD Animal Health, Deventer, The Netherlands

10:00-10:15 **1.2 FAILURE TO DETECT M. AVIUM SUBSPECIES PARATUBERCULOSIS BY FLUORESCENT IN SITU HYBRIDIZATION (FISH) IN JOHNE'S OR CROHN'S DISEASE USING A PROPRIETARY ASSAY.** Robert Greenstein[1], Liya Su[1], Peter Fam[1], Judith Stabel[2], Sheldon Brown[1]. 31
[1]JJP Veteran Affairs Medical Center Bronx NY USA
[2]USDA-ARS-NADC Ames Iowa USA

10:15-10:30 **1.3 DIVERSE HISTOPATHOLOGICAL AND MICROBIOLOGICAL FINDINGS IN EXPERIMENTAL MYCOBACTERIUM AVIUM SUBSPECIES PARATUBERCULOSIS INFECTION IN CATTLE.** Richard Whittington[1], Douglas Begg[1], Karren Plain[1], Kumudika de Silva[1], Ratna Gurung[1], Alison Gunn[1], Auriol Purdie[1]. 32
1 The University of Sydney

10:30-10:45 **1.4 PREDICTION OF THE GENETIC SUSCEPTIBILITY TO PARATUBERCULOSIS IN DAIRY CATTLE USING ALLELIC COMBINATIONS OF FIVE SINGLE NUCLEOTIDE POLYMORPHISMS IN CD209, SLC11A1, SP110, AND TLR2 GENES.** María Canive[1], Rosa Casais[2], Patricia Vázquez[1], Ana Balseiro[2], José M. Prieto[2], Javier Amado[3], José A. Jiménez[4], Joseba Garrido[1], Ramón A. Juste[1,5], Marta Alonso-Hearn [1]. 33
[1]NEIKER- Instituto Vasco de Investigación y Desarrollo Agrario, Animal Health Department, Derio, Bizkaia, Spain
[2]SERIDA, Servicio Regional de Investigación y Desarrollo Agroalimentario. Centro de Biotecnología, Deva, Asturias, Spain
[3]LSAPA, Animal Health Laboratory of the Principality of Asturias, Department of Microbiology, Gijón, Asturias, Spain

[4]CONAFE, Spanish Federation of Holstein Cattle, Madrid, Spain
[5]SERIDA, Regional Service of Agri-Food Research and Development, Villaviciosa, Asturias, Spain

10:45-11:00 **1.5 EARLY DIAGNOSIS OF PARATUBERCULOSIS INFECTION USING ELISAS BASED ON THE DETECTION OF HOST BIOMARKERS.** Rosa Casais[1], Cristina Blanco[1], Ana Balseiro[1], Ramón Antonio Juste[1], Begoña Menendez[1], José Miguel Prieto[1], Rosana Torremocha[3], Beatriz Soriano[4], Ricardo Ramos[3], Carlos Llorens[4], Francisca Copano[5], Javier Amado[5], Maria Canive[6], Marta Alonso[6] 34

[1]SERIDA, Servicio Regional de Investigación y Desarrollo Agroalimentario. Centro de Biotecnología Animal, Deva, Asturias, Spain
[2]SERIDA, Regional Service of Agri-Food Research and Development, Villaviciosa, Asturias, Spain
[3]Parque Científico de Madrid, Unidad de Genómica, Campus de Cantoblanco, Madrid
[4]Biotechvana, Paterna, Valencia, Spain
[5]LSAPA, Animal Health Laboratory of the Principality of Asturias. Departament of Microbiology, Gijón, Asturias, Spain
[6]NEIKER- Instituto Vasco de Investigación y Desarrollo Agrario, Animal Health Department, Derio, Bizkaia, Spain

11:00-11:15 **1.6 CHARACTERIZING RESPONSES OF IMMUNE CELL SUBSETS FROM M. PARATUBERCULOSIS (MAP) TEST POSITIVE AND TEST NEGATIVE COWS FROM COMMERCIAL HERDS TO MAP ANTIGEN STIMULATION IN VITRO.** Paul Coussens, Meredith C. Frie 35

11:15-11:30 **1.7 COMPARISON OF SHEEP, GOATS, AND CALVES AS INFECTION MODELS FOR MYCOBACTERIUM AVIUM SUBSP. PARATUBERCULOSIS**. Judith Stabel[1], John P. Bannantine[1], Jesse M. Hostetter[2]. 36

[1]USDA-ARS-National Animal Disease Center
[2]Iowa State University

11:30-11:45 **1.8 LDLR FACILITATES MYCOBACTERIAL INFECTION-INDUCED CHANGES TO HOST LIPID METABOLISM.** Matt D. Johansen[1], Auriol C. Purdie[1], Kumudika de Silva[1], Karren M. Plain[1], Stefan H. Oehlers[2]. 37

[1]University of Sydney
[2]Centenary Institute

11:45-12:00 **1.9 EFFECTS OF VACCINATION ROUTE ON THE IMMUNE RESPONSE IN THE RABBIT PARATUBERCULOSIS INFECTION MODEL.** Iraia Ladero[1], Rakel Arrazuria[1], Elena Molina[1], Miguel Fernández[2], Marcos Royo[2], Joseba Garrido[1], Ramón Juste[1], Valentín Pérez[2], **Natalia Elguezabal[1]**. 38

[1]NEIKER-Instituto Vasco de Investigación y Desarrollo Agrario, Animal Health Department, Derio, Bizkaia, Spain
[2]Departamento de Sanidad Animal, Instituto de Ganadería de Montaña (CSIC-Universidad de León), Facultad de Veterinaria, C/ Profesor Pedro Cármenes s/n, 24071 León

12:00-12:15 **1.10 CYTOKINE GENE EXPRESSION PROFILES IN THE BLOOD OF CLINICAL AND NON-CLINICAL SHEEP AND GOATS INFECTED WITH MYCOBACTERIUM AVIUM SUBSPECIES PARATUBERCULOSIS.** Ganesh Gangaram Sonawane[1], Bhupendra Nath Tripathi[1], L.L.L. Prince[2], Rajendra Kumar Meena[1]. 39

1 ICAR-National Research Centre on Equines, Hisar-125001(Haryana), India

2 ICAR - Directorate of Poultry Research, Rajendranagar, Hyderabad 500 030 (Telangana), India

12:15-12:30 **1.11 CONTROL OF JOHNE'S DISEASE IN A NEW ZEALAND DAIRY HERD USING TEST AND CULLING.** Frank Griffin[1], Rory O´Brien[1], Simon Liggett[1], Andrew Bates[2]. 40

1 Disease Research Limited, Invermay Agricultural Centre, Mosgiel, New Zealand

2 Vetlife, Centre for Dairy Excellence, Geraldine, New Zealand

12:30-12:45 **1.12 RECOMBINANT SECRETARY PROTEINS BASED 'COCKTAIL ELISA' TO DIFFERENTIATE BETWEEN INFECTED AND VACCINATED BOVINES AGAINST MYCOBACTERIUM AVIUM SUBSPECIES PARATUBERCULOSIS.** Kundan Kumar Chaubey[1], Shoor Vir Singh[1], Rinkoo Devi Gupta[2], Kirti Dua[3], Shukriti Sharma[3], Niranjana Sahoo[4]. 41

[1]Animal Health Division, Central Institute for Research on Goats, Makhdoom, PO- Farah, Mathura, UP, India

[2]Department of Life Sciences and Biotechnology, South Asian University, New Delhi, India

[3]Department of Epidemiology and Preventive Medicine, College of Veterinary Science and Animal Husbandry, GADVASU, Ludhiana, Punjab, India

[4]Department of Epidemiology and Preventive Medicine, College of Veterinary Science and Animal Husbandry, OUAT, Bhubaneswar, Odisha, India

13:00-14:45 Lunch time.

2nd Oral Session. Moderators: Natalia Elguezabal & David Kelton (14:45-16:00)

14:45-15:00 **1.13 IMMUNE RESPONSE INDUCED BY A LOCAL STRAIN OF MYCOBACTERIUM AVIUM SUBSP. PARATUBERCULOSIS IN ANIMAL MODELS**. María Alejandra Colombatti Olivieri[1], Roberto Damian Moyano[1], María José Gravisaco[1], Gabriel Eduardo Traveria[2], María de la Paz Santangelo[1], María Isabel Romano[1]. 42

[1]Instituto de Biotecnología, INTA Castelar, Buenos Aires, Argentina

[2]Centro de Diagnostico e Investigaciones Veterinarias (CEDIVE) de la Facultad de Ciencias Veterinarias, Universidad de La Plata, Buenos Aires, Argentina

15:00-15:15 **1.14 MAP-SPECIFIC VOLATILE ORGANIC COMPOUND PROFILE: A COMPARATIVE ANALYSIS OF THREE DIFFERENT IN VITRO STUDIES**. Anne Küntzel[1], Sina Fischer[1], Andreas Bergman[2], Peter Oertel[2], Philip Trefz[2], Wolfram Miekisch[2], Jochen Schubert[2], Petra Reinhold[1], Heike Köhler[1.] 43

[1]Friedrich-Loeffler-Institut, Institute of Molecular Pathogenesis, Germany

[2]University of Rostock, Department of Anesthesia and Intensive Care, Germany

15:15-15:30 **1.15 EX VIVO VACCINATION WITH A PLGA/MPLA NANOPARTICLE VECTORED A 35 KDA MAJOR MEMBRANE PROTEIN FROM MYCOBACTERIUM AVIUM PARATUBERCULOSIS ELICITS KILLING OF INTRACELLULAR BACTERIA BY CD8 T CELLS.** Cleverson D Souza[1], Gaber S. Abdellrazeq[1], Mahammoud M. Elnaggar[2], John P. Bannantine[3], Julianne Hwang[4], William C. Davis[2]. 44

[1]Washington State University, Faculty Veterinary Medicine, Alexandria University, Egypt
[2]Vet Micro/Pathol, Washington State University, Pullman, WA, U.S., Faculty Veterinary Medicine, Alexandria University, Egypt
[3]National Animal Disease Center, USDA-Agricultural Research Service, Ames, US
[4]Veterinary Clinical Sciences, Washington State University, Pullman, US

15:30-15:45 **1.16 DISEASE STATE INFLUENCES THE PRESENCE OF MACROPHAGES AND MYCOBACTERIUM AVIUM SUBSP. PARATUBERCULOSIS IN BOVINE INTESTINAL TISSUE.** Caitlin Jenvey[1], Jesse M. Hostetter[1], Adrienne L. Shircliff[2], John P. Bannantine[2], Judith R. Stabel[2]. 45

[1]Iowa State University
[2]USDA-ARS-National Animal Disease Center

15:45-16:00 **1.17 MACROPHAGE ANALYSIS TO IDENTIFY FUNCTIONAL GENETIC MARKERS ASSOCIATED WITH JOHNE'S DISEASE.** Nathalie Bissonnette[1], Jean-Simon Brouard[1], Olivier Ariel[1], Eveline Ibeagha-Awemu[1], Filippo Miglior[2]. 46

[1]Agriculture and Agri-Food Canada, Sherbrooke Research and Development Centre, Sherbrooke, QC, Canada
[2]Centre for Genetic Improvement of Livestock, University of Guelph, Canadian Dairy Network, Guelph, ON, Canada

16:00-16:35 Coffee break

16:35-19:00 Poster Session 45

WEDNESDAY JUNE 6th

Session 2. Map Control Programs and Epidemiology.

8:00- 8:45 **Lecture. Dr. Richard Whittington (Australia): "INTERNATIONAL CONTROL PROGRAMS FOR PARATUBERCULOSIS WHO, HOW AND WHY?"** 93

Moderators: Victoria E. Castrellón Ahumada & Miguel Salgado

1st Oral Session. Moderators: Judy Stabel & Joseba Garrido (8:45-10:45)

8:45- 9:00 **2.1 ASSIGNMENT OF MAP TYPE C STRAINS FROM GERMANY TO GLOBAL PHYLOGENETIC MAP GROUPS.** Petra Moebius[1], Heike Koehler[1]. 94
[1]Institute of Molecular Pathogenesis, Friedrich-Loeffler-Institut, Jena.

9:00- 9:15 **2.2 BREEDING FOR RESISTANCE AGAINST PARATUBERCULOSIS: HIGH GENETIC CORRELATION BETWEEN ANTIBODY RESPONSE AND FAECAL SHEDDING**. Lydia C.M. de Haer[2], Gerben de Jong[1], Maarten F. Weber[2]. 95
[1]CRV Arnhem, the Netherlands
[2]GD Animal Health, Deventer, the Netherlands.

9:15- 9:30 **2.3 ECONOMIC IMPACT OF CONTROL OPTIONS FOR JOHNE'S DISEASE IN CANADA**. Philip Rasmussen[1], Zhaoxue Ci[1], David Hall[1]. 96
[1]University of Calgary.

9:30-9:45 **2.4 AN EFFECTIVE CONTROL PROGRAM USING A POOLED FAECAL REAL-TIME PCR ASSAY IN HERDS WITH JOHNE'S DISEASE**. Satoko Kawaji[1], Reiko Nagata[1], Akiko Mita[2], Makoto Osaki[1], Yasuyuki Mori[1]. 97
[1]National Institute of Animal Health, NARO
[2]National Livestock Breeding Center

9:45-10:00 **2.5 A WEB BASED PARATUBERCULOSIS RISK ASSESSMENT AND MANAGEMENT SYSTEM FOR DAIRY FARMERS.** Dick Sibley BVsc HonFRCVS[1], Pete Orpin BVSc MRCVS[1]. 98
[1]Park Vet Group

10:00-10:15.**2.6 FIELD VALIDATION OF DIVA ASSAY FOR INACTIVATED PARATUBERCULOSIS VACCINE**. Sujata Jayaraman[1], Mukta Jain[1], Kundan Kumar Chaubey[2], S. V. Singh[2], G. K. Aseri[1], Jagdip Singh Sohal[3]. 99
[1]Amity Institute of Microbial Technology, Amity University Rajasthan, Jaipur, India.
[2]Microbiology Laboratory, Central Institute for Research on Goats, Mathura, India
[3]Amity Center for Mycobacterial Disease Research, Amity University Rajasthan, Jaipur, India

10:15- 10:30 **2.7 PROTECTIVE LIVE ATTENUATED AND NANO-VACCINES AGAINST JOHNE'S DISEASE.** Adel M. Talaat[1], Akanksha Thukral[1], Chungyi Hansen[1], Kathleen Ross[2], Balaji Narasimhan[2]. 100
[1]University of Wisconsin-Madison.
[2]Iowa State University

10:30-10:45 **2.8 MYCOBACTERIUM AVIUM SSP. PARATUBERCULOSIS (MAP) MOLECULAR DIVERSITY IN LATIN AMERICA AND THE CARIBBEAN: A SYSTEMATIC REVIEW**. Nathalia M Correa-Valencia[1], Miguel Hernández-Agudelo[1], Jorge A Fernández-Silva[1]. 101
[1]Centauro, Escuela de Medicina Veterinaria, Facultad de Ciencias Agrarias, Universidad de Antioquia, Colombia

10:45-11:00 Coffee break

2nd Oral Session. Moderators: Vivek Kapur & Nicola Pozzato (11:00-12:15)

11:00 -11:15 **2.9 PHYLOGENETIC ANALYSIS AND EPIDEMIOLOGIC MODELING OF THE EFFECT OF MYCOBACTERIUM AVIUM SUBSP. PARATUBERCULOSIS GENOTYPE ON MILK PRODUCTION ON THREE MINNESOTA DAIRY FARMS.** Jonah Cullen[1], Yuanyuan Wang[1], Juan Abrahante[1], Fernanda Shoyama[2], Srinand Sreevatsan[2], Scott Wells[1] 102
[1]University of Minnesota.
[2]Michigan State University

11:15 -11:30 **2.10 LONG TERM RESULTS OF AN EXPERIMENTAL VACCINATION TRIAL IN DAIRY CATTLE.** Joseba M. Garrido[1], Natalia Elguezabal, Ramón A. Juste[2], Miriam Serrano, María V. Geijo, Elena Molina, Iker A. Sevilla. 103
[1]Neiker-Instituto Vasco de Investigación y Desarrollo Agrario. Departamento de Sanidad Animal, Derio, Bizkaia, Spain.
[2]Serida, Servicio Regional de Investigación y Desarrollo Agroalimentario de Asturias, Villaviciosa, Asturias, Spain.

11:30 – 11:45 **2.11 ARE CATTLE INFECTED WITH MULTIPLE STRAINS OF MAP? A NEW COMPUTIONAL METHOD TO DETECT FROM WHOLE GENOME SEQUENCING DATA.** Yuanyuan Wang[1], Scott J. Wells[1]. 104
[1]University of Minnesota, Twin Cities

11:45-12:00 **2.12 MYCOBACTERIUM AVIUM SUBSP. PARATUBERCULOSIS DETECTION IN WILD RABBITS FROM SOUTHERN SPAIN.** Natalia Elguezabal[1], Elena Molina[1], Rakel Arrazuria[1], Iker A Sevilla[1], Mariví Geijo[1], Christian Gortazar[2], Miguel Fuertes[3], Valentín Pérez[3], Ramón Juste[1], Joseba M Garrido[1]. 105
[1]NEIKER-Instituto Vasco de Investigación y Desarrollo Agrario, Animal Health Department, Derio, Bizkaia, Spain.
[2]SaBio, Instituto de Investigación en Recursos Cinegéticos IREC (CSIC-UCLM-JCCM), Ciudad Real, Spain.
[3]Department of Animal Health, Faculty of Veterinary Medicine, University of León, Spain

12:00 – 12:15 **2.13 PRACTICAL EXPERIENCES IN DELIVERING A COMMERCIALLY DRIVEN NATIONAL JD PROGRAM IN THE UNITED KINGDOM.** Pete Orpin[1], Richard Sibley[1]. 106
[1]Westridge Veterinary Group

12:15 – 14:00 Poster session 107

14:00– 15:00 Lunch

15:00 Afternoon Free

THURSDAY. JUNE 7th

Session 3. Pathogenomics and Map biology.

8:00- 10:10 Foro Iberoamericano (By invitation)

10:30-11:15 **Lecture. Dr Jeroen De Buck (Canada): "RECENT DEVELOPMENTS IN PATHOGENOMICS TO UNDERSTAND MAP (*MYCOBACTERIUM AVIUM SUBSP. PARATUBERCULOSIS*) PATHOGENESIS AND DIAGNOSE JOHNE'S DISEASE".** 128

11:15-11:30 Coffee break

Moderators: Victoria E. Castrellón Ahumada & Miguel Salgado

Oral Session. Moderators: Karen Stevenson & Adel Talaat (11:30-13:00)

11:30- 11:45 **3.1 ON THE USE OF WHOLE GENOME SEQUENCING TO UNVEIL LOCAL SPREAD OF A PARATUBERCULOSIS CLONE WITHIN A SINGLE HERD**. Matteo Ricchi[1], Simone Russo[1], Erika Scaltriti[2], Simone Leo[1], Luca Bolzoni[2], Norma Arrigoni[1] 131

[1]Istituto Zooprofilattico Sperimentale della Lombardia e dell'Emilia Romagna, National Reference Centre for paratuberculosis, Strada Faggiola 1, Podenzano (PC), Italy.

Istituto Zooprofilattico Sperimentale della Lombardia e dell'Emilia Romagna, Risk Analysis Unit, Via dei Mercati 13/A, Parma 43126, Italy.

11:45- 12:00 **3.2 EVALUATION OF A MAP MUTANT IN BOVINE BMDM AND IN A MOUSE MODEL, AND NEW METHODS FOR M. AVIUM SUBSPECIE PARATUBERCULOSIS MUTAGENESIS.** María de la Paz Santangelo[1], Mariana Viale[1], Alejandra Colombatti[1], Natalia Alonso [1], William Davis[2], María Isabel Romano[1]. 132

[1]Instituto de Biotecnología, INTA. Buenos Aires, Argentina. CONICET

[2]Department of Veterinary Microbiology and Pathology, College of Veterinary Medicine, Washington State University, Pullman, WA 99164, USA

12:00-12:15 **3.3 MYCOBACTERIUM AVIUM SUBSP. PARATUBERCULOSIS: IN VITRO INTERACTION WITH AND BOVINE SPERM CELLS.** Isis Espeschit[1], Ana Carolina Silva Faria[1], Daniel Mendonça de Araújo Lima[1], Breno Soares Camilo[1], David Germano Gonçalves Schwars[1], Mariana Barros[1], Sanely Lourenço da Costa[1], Maria Aparecida Scatamburlo Moreira[1]. 133

[1]Universidade Federal de Viçosa.

12:15-12:30 **3.4 ASSESSMENT OF MYCOBACTERIUM AVIUM SUBSP. PARATUBERCULOSIS VIABILITY USING FLOW CYTOMETRY.** Timothy Scott[1], Marina Buciuc[1]. 134

[1]Minnesota State University, Mankato

12:30-12:45 **3.5 TRANSCRIPTOME ANALYSIS OF ILEUM TISSUES FROM MYCOBACTERIUM AVIUM SSP. PARATUBERCULOSIS INFECTED COWS REVEALS IMPORTANT GENES AND PATHWAYS FOR JOHNE'S DISEASE.** 135
Nathalie Bissonnette[1], Duy N. Do[1], Pier-Luc Dudemaine[2], Eveline M. Ibeagha-Awemu[1].
[1]Agriculture and Agri-Food Canada, Sherbrooke Research and Development Centre, Sherbrooke, Qc, Canada
[2]Agriculture and Agri-Food Canada, Sherbrooke Research and Development Centre, Sherbrooke, Department of Animal Science, McGill University, Qc, Canada.

12:45-13:00 **3.6 HOST GENE EXPRESSION SIGNATURE OF IMMUNE-REGULATORY GENES IN WHOLE BLOOD OF CATTLE WITH SUBCLINICAL INFECTION OF MYCOBACTERIUM AVIUM SUBSP. PARATUBERCULOSIS**. Han Sang Yoo[1], Hyun-Eui Park[1], Hongtae Park[1], Young Hoon Jung[2]. 136
[1]Department of Infectious Diseases, College of Veterinary Medicine, Seoul National University, Korea.
[2]Department of Animal Resources Development, National Insitute of Animal Science, Rural Development Administyration, Korea

13:00-14:30 Lunch time

14:30-15:15 **Lecture. Ramón A. Juste Jordan (Spain): "Chronic regional intestinal inflammatory disease: A walk through the species".** 130

15:15-15:30 Coffe break

15:30- 16:10 "Richard S. Merkal Award"
Moderator: Scott Wells

DETECTION OF MARKER-SPECIFIC IMMUNE RESPONSES IN CALVES AGAINST A NOVEL MARKED MAP PARENT VACCINE STRAIN Lucy Luo[1], Jeroen De Buck[1]. 156
[1]University of Calgary

PREVALENCE OF MYCOBACTERIUM AVIUM SUBSP. PARATUBERCULOSIS ACROSS CANADA, BASED ON ENVIRONMENTAL SAMPLING. Caroline S. Corbett[1], S. Ali Naqvi[1], Cathy Bauman[2] Jeroen De Buck[1], Karin Orsel[1], David F. Kelton[2], Herman W. Barkema. 157
[1]University of Calgary
[2]University of Guelph

16:10 -17:00 Poster session 137

17:00-18:00 General Meeting

18:00-18:15 Group Photography

FRIDAY JUNE 8th

Session 4. Public Health and Map Environment.

8:00- 8:45 Lecture. Dr. Rodrick Chiodini (USA): "Crohn's Disease and Johne's disease: Why everyone doesn't believe you" 143

Moderators: Victoria E. Castrellón Ahumada & Miguel Salgado

Oral Session. Moderators: Tim Bull & Robert Greenstein (8:45-10:00)

8:45-9:00 **4.1. NOVEL NANO-IMMUNO-TEST FOR THE DETECTION OF LIVE PARATB BACILLI IN THE MILK OF DOMESTIC LIVESTOCK.** Shoor Vir Singh[1], Manju Singh[1], Bjorn John Stephan[2], Manali Dutta[2], Gajendra Kumar Aseri[2], Jagdip Singh Sohal[2]. 144
[1]Animal Health Division, Central Institute for Research on Goats, Makhdoom, PO-Farah- 281 122, Dist. - Mathura, Uttar Pradesh, India.
[2]AIMT & AIB, Amity University Rajasthan, Jaipur, India.

9:00-9:15 **4.2 ESTIMATION OF PERFORMANCE CHARACTERISTICS OF ANALYTICAL METHODS FOR MYCOBACTERIUM AVIUM PARATUBERCULOSIS (MAP) DETECTION IN MILK.** Butot Sophie, Ricchi Matteo[1], Sevilla Iker[2], Michot Lise[3], Molina Elena[2], Tello Maitane[2], Russo Simone[1], Tomas David[3]. 145
[1]IZSLER
[2]NEIKER
[3]Nestlé Research Center

9:15- 9:30 **4.3 EFFECT OF CALCIUM HYDROGEN CARBONATE MESOSCOPIC CRYSTALS TO MYCOBACTERIUM AVIUM SUBSP. PARATUBERCULOSIS.** Eiichi Momotani, Rumiko Onishi, Koichi Furusaki, Takashi Onodera 146

9:30-9:45 **4.4 FREE-LIVING AMOEBAE AS AN ENVIRONMENTAL HOST FOR MYCOBACTERIUM AVIUM SUBSP. PARATUBERCULOSIS.** Hechard Yann[1], Ascel Samba-Louaka[1], Etiene Robino[1], Thierry Cochard[2], Willy Aucher[1], Franck Biet[2]. 147
[1]University of Poitiers
[2]INRA Nouzilly

9:45-10:00 **4.5 MYCOBACTERIUM AVIUM SUBSPECIES PARATUBERCULOSIS AND MYELIN BASIC PROTEIN SPECIFIC EPITOPES ARE HIGHLY RECOGNIZED BY SERA FROM PATIENTS WITH NEUROMYELITIS OPTICA SPECTRUM DISORDER.** Marco Bo[1], Giannina Arru[1], Magdalena Niegowska[2], Gianpietro Sehi[1], Leonardo Sechi[2]. 148
[1]Department of Clinical and Experimental Medicine, University of Sassari, Italy.
[2]Department of Biomedical Sciences, University of Sassari, Sassari, Viale San Pietro 43b, Italy

10:00-10:15 Coffee break

10:15-11:15 Poster sesión 149

11:15-12:15 Synopsis of the Colloquium, next 15th ICP.

Abstracts

LECTURE: ASPECTS OF PATHOLOGY AND PATHOGENESIS OF DOMESTIC RUMINANT PARATUBERCULOSIS.

CLAUS D. BUERGELT, DVM, PhD, DACVP
Professor Emeritus
University of Florida
Gainesville, Florida 32610 USA

INTRODUCTION
Paratuberculosis (Johne's disease) caused by *Mycobacterium avium* subsp. *paratuberculosis* (MAP) has been primarily described in both domestic and wild ruminants. It is distributed around the world where it has significant economic and welfare impacts on meat and milk production. Clinically the disease is characterized by profuse diarrhea, weight loss and dehydration in adult animals. Diarrhea is not constant findings in small ruminants. The small intestinal tract is the target organ for infection with MAP. Various clinical, morphologic and immunologic stages have been described during the infection phase. Pathologic changes in ruminant paratuberculosis are characteristic and fairly consistent, though subtle differences exist between domestic ruminant species. Early events of pathogenesis involve intestinal epithelial attachment and invasion followed by intracellular trafficking of MAP in adjacent macrophages. The understanding of these early events may be useful for therapeutic intervention procedures to prevent infection to unfold and to transgress into disease.

PATHOLOGY
Lesions are most common in the upper small intestinal tract (ileum) and mesenteric lymph nodes. Characteristic **macroscopic** changes of clinically diseased animals include (1) unstretchable rugal mucosal corrugation; (2) lymphadenomegaly of mesenteric (ileo-cecal) lymph nodes with coalescing areas of white-yellow caseous exudate and mineralization in goats; (3) cording and beading of subserosal and mesenteric lymphatic vessels (lymphangitis). In some instances, particularly in small ruminants, mucosal corrugation may be absent in clinical cases.

Microscopically, the typical clinical case of paratuberculosis is characterized by a non-caseating granulomatous inflammation within the intestinal lamina propria and submucosa composed predominantly of epithelioid macrophages and inflammatory giant cells of Langhans type. In subclinical paratuberculosis with mild lesions the Langhans type giant cell may the early inflammatory cell to occupy the lamina propria and its presence may be helpful for the presumptive diagnosis of Johne's disease. Acid fast stains usually demonstrate a plethora of engulfed bacilli within epithelioid macrophages (multibacillary type) and to a significant lesser degree in Langhans type giant cells. Secondary dilation of villous lacteals from lymphatic inflammatory obstruction leads to protein-losing enteropathy.
Extraintestinal sites for infection by MAP, mainly detected by microscopy as microgranulomas, have been described in the liver, kidney, lung and udder.

In some **cattle** with clinical disease and occasionally in goats, unrelated lesions diagnosed as focal mineralization have been described in the intima of the ascending aorta and endocardium of the cardiac left atrium. Investigations into the pathogenesis are lacking in Johne's disease, but comparative studies in noncaseating granulomatous diseases such as human sarcoidosis have incriminated activated macrophages to metabolize 25-hydroxy vitamin D_3 to1, 25 dihydroxy vitamin D_3 as a side product.

In **goats**, caseation with calcification is seen in infected mesenteric lymph nodes. A recent publication described a granulomatous vasculitis in medium-sized arteries in the intestinal submucosa and in the paracortex of mesenteric lymph nodes in goats experimentally infected with MAP. In some of these experimental animals ulceration of the intestinal mucosa over Peyers patches had occurred. Intestinal strictures and fibrous adhesions in naturally infected goats have been described in Norway.
In **sheep** with clinical disease 15% of the cases have no gross lesions in the intestine. Other clinical cases have extreme thickened intestinal walls. Some infected sheep exhibit yellow pigmentation of the intestinal mucosa.

INITIAL EVENTS OF INTESTINAL CELL-MAP INTERACTION

Adherence and internalization of bacilli are major early events in the process of MAP infection. The adhesive glycoprotein fibronectin has been shown to play an integral part in these mechanisms. Fibronectin can mediate the adherence of MAP to the host cell by interacting with specific binding sites on both bacilli and host cells. Fibronectin attachment proteins (FAP) are a family of fibronectin (FN)-binding proteins present on mycobacteria. Beta-1 integrins function as host cell receptors for FN-opsonized mycobacteria to build a fibronectin bridge for bacterial internalization. M cells (microfold cells) are unique intestinal epithelial cells in the dome epithelium overlying the subepithelial lymphoid follicles displaying beta-1 integrins on their luminal surface in high density. Therapeutic disruption of FN-binding may diminish invasion of MAP into M cells. M cells represent an important pathway for the direct transport of intestinal antigens to the gut-associated lymphoid tissue (GALT). Experimental calf gut-loop studies by Momotani et al. have shown the initial uptake of MAP by M cells to breach the mucosal barrier to reach antigen-processing denditric (Langerhans) cells and subepithelial macrophages associated with GALT. Invasion of non-M cell villous enterocytes also can occur through other epithelial routes and can follow FN-independent mechanisms.

After transcytosis through microfold epithelial cells (M cells) or transcellular and paracellular routes in intestinal enterocytes MAP is taken up by subepithelial macrophages in the lamina propria and gut-associated lymphoid tissue (GALT). Macrophages are the target cells for MAP to survive and to multiply intracellularly. Invaded macrophages are carried to regional lymph nodes from intestinal sites. MAP can also be picked up by resident dendritic cells. Monocytes and lymphocytes are recruited to the site of infection and become activated to release a variety of cytokines. The intracellular fate of MAP in macrophages is governed by various pathways that after phagosomal incorporation and phago-lysosome fusion include replication pathway in bacteriophorous vacuole, antigen processing pathway for partial degradation and exocytotic pathway after complete degradation. Intracellullar MAP successfully have developed strategies to escape an organized host defense system. These strategies include 1) escape from the phagosome into the cytoplasm, 2) avoidance of phagosomal maturation and phagosome-lysosome fusion, 3) modification of lysosomal contents, and 4) passive protection by the cell wall envelope. The ability to reduce apoptosis of infected macrophages is an alternative mechanism to evade host defense as is interference with autophagy during infection to avoid restriction of replication.

REFERENCES

Adams, JS, Gacad, MA: Characterization of 1alpha- Hydroxylation of Vitamin D Sterols by Cultured Alveolar Macrophages from Patients with Sarcoidosis. J.Exp.Med. 161:755-765, 1985

Bennantine, JP, Bermudez, LE: No Holes Barred: Invasion of the Intestinal Mucosa by *Mycobacterium avium* subsp. *paratuberculosis.* Infect.Immunity 81:3960-3965

Clarke, CJ: The Pathology and Pathogenesis of Paratuberculosis in Ruminants and Other Species. J.Comp.Path. 116:217-26, 1997

Hines, SA, Buergelt, CD, Wilson, JH, Bliss, EL: Disseminated *M. paratuberculosis* Infection in a Cow. JAVMA 190:6811-83, 1987

Kinsella, Rl, Nehls, EM, Stallings, Ch L: Roles of Autophagy Proteins in Immunity and Host Defense. DOI:10.11777/0300985818754967

Krueger, C, Koehler,H, Liebler-Tenorio,EM: Sequential Development of Lesions 3, 6,9 and 12 Months After Experimental Infection of Goat Kids with *Mycobacterium avium* subsp *paratuberculosis*. Vet. Pathol. 52:276-290, 2015

Momotani, E, Whipple,DL,Thiermann,AB, Cheville,NF: Role of M Cells and Macrophages in the Entrance of *Mycobacterium paratuberculosis* into Domes of Ileal Peyers Patches in Calves. Vet. Pathol. 25:131-137,1988

Sivakumar Petriasamy, Bhupendra N. Tripathi, Nem Singh: Mechanisms of *Mycobacterium avium* subsp. *paratuberculosis* induced apoptosis and necrosis in bovine macrophages. Vet.Micro. 165:392-401,2013

Secott,TE, Lin, TL, Wu,CC: Fibronectin Attachment Protein Homologue Mediates Fibronectin Binding by *Mycobacterium avium* subsp. *paratuberculosis.* Infect.Immunity 69:2075-2082, 2001

Tessema, MZ, Koets, AP, Rutten, VPMG, Gruys, E: HOW DOES *MYCOBACTERIUM AVIUM* SUBSP. *PARATUBERCULOSIS* RESIST INTRACELULAR DEGRADATION? Vet.Quart. 223:153-162, 2001

Oral

1.1 LIMITED ADAPTIVE IMMUNE RESPONSES OF GOAT KIDS TO MAP BACTERIN VACCINATION WHEN RAISED IN A MYCOBACTERIUM AVIUM SSP PARATUBERCULOSIS (MAP) FREE ENVIRONMENT.

Ad Koets[1], Lars Ravesloot[1], Robin Ruuls[1], Annemieke Dinkla[1], Karianne Lievaart-Peterson[2].

[1] *Wageningen Bioveterinary Research, Lelystad, The Netherlands*

[2] *GD Animal Health, Deventer, The Netherlands*

Vaccination is the principal control strategy for caprine paratuberculosis in the Netherlands. Most goat dairy farms with endemic paratuberculosis systematically vaccinate goat kids in the first month of life with a commercially available whole cell MAP vaccine. A single vaccination of goat kids provides protection against development of clinical paratuberculosis.

We hypothesized that the development of sustained adaptive immune responses in goats vaccinated once at young age may, at least partially, be dependent on the environment they are raised in.

We sourced 24 female twins from a MAP unsuspected non-vaccinated herd and raised them in a MAP free environment. We vaccinated one twin goat kid of each pair resulting in a comparative study with 24 vaccinated and 24 control goat kids. These goats were sampled at regular intervals during the first 6 months of live. We evaluated the early B and T cell immune response to vaccination in these young goat kids. In comparison goats born in the same year but vaccinated and raised on MAP endemic farms are being investigated.

Results indicated that adaptive immune responses to vaccination are limited in a MAP free environment. Both MAP antigen specific antibody responses as well as antigen specific interferon-gamma responses peak at 4-5 weeks post vaccination but wane rapidly thereafter. Preliminary data from age matched goats vaccinated and raised in a MAP endemic environment show high antibody responses.

These data support a role for environmental exposure to MAP in sustaining adaptive immune responses to vaccination.

1.2 FAILURE TO DETECT M. AVIUM SUBSPECIES PARATUBERCULOSIS BY FLUORESCENT IN SITU HYBRIDIZATION (FISH) IN JOHNE'S OR CROHN'S DISEASE USING A PROPRIETARY ASSAY.

Robert Greenstein[1], Liya Su[1], Peter Fam[1], Judith Stabel[2], Sheldon Brown[1].

[1]*JJP Veteran Affairs Medical Center Bronx NY USA*
[2]*USDA-ARS-NADC Ames Iowa USA*

M. avium subspecies *paratuberculosis* (MAP) causes Johne's disease. MAP may be zoonotic, responsible for, at a minimum, Crohn's disease. Thus, the presence or absence of MAP needs to be reliably diagnosed. The "gold standard" for detection of MAP in cattle is culture, followed by DNA sequencing. Culture of MAP in Crohn's has been achieved in very few laboratories.

The purpose of this study was to evaluate a proprietary (Affymetrix RNA view®) FISH assay for MAP RNA in humans with Crohn's, using Johne's tissue as the positive control.

Intestine from steer with Johne's disease and humans with Crohn's were assayed according to Affymetrix's instructions. Published genomes were used to custom design probes. IS900 for MAP and bovine and human - actin (as species specific eukaryotic housekeeping controls). We attempted to prevent false positive signal in the "no- probe" control, by modifying wash solutions, using recommended and derivative hydrochloric acid titration and different fluorescent filters (TritC for Texas Red and "Hope" for Cy-5.)

Repetitively, false positive signal was observed in our "no-probe" negative control. Attempts to correct this according to the manufacturers suggestions, and with multiple derivative techniques have been unsuccessful.

We performed the Affymetrix RNA view® according to the manufacture's instruction and used multiple variations on the manufacture's recommended suggestions to correct for false positive signal. Repetitively, "no-probe" positive controls indicate that this assay cannot be used to reliably detect MAP in pre-frozen intestine of cattle with Johne's disease. Nor can it be used in the possible identification of MAP in frozen tissue from humans with Crohn's disease.

1.3 DIVERSE HISTOPATHOLOGICAL AND MICROBIOLOGICAL FINDINGS IN EXPERIMENTAL MYCOBACTERIUM AVIUM SUBSPECIES PARATUBERCULOSIS INFECTION IN CATTLE.

Richard Whittington[1], Douglas Begg[1], Karren Plain[1], Kumudika de Silva[1], Ratna Gurung[1], Alison Gunn[1], Auriol Purdie[1].

[1]*The University of Sydney*

The aim of this study was to evaluate histopathological and microbiological findings among calves following experimental infection with a low dose of *Mycobacterium avium* subspecies *paratuberculosis* (MAP).

Twenty calves were followed for more than 4.5 years after oral exposure to MAP; intestinal biopsies were performed at two time points up to 26 months and there was a thorough necropsy.

Commencing 9 weeks post infection, 15 calves shed viable MAP in faeces, 9 on more than one occasion and 2 became persistent shedders, associated with gut pathology. None of the cattle developed clinical signs of Johne's disease. At necropsy, the tissues of 7 cattle were infected with MAP. In 3 of these multiple tissues contained viable MAP, while in 4 MAP was recovered only from a mesenteric lymph node. Two cattle had histopathological lesions of Johne's disease at the time of necropsy and these also had disseminated infections with MAP in the hepatic lymph node or liver. Based on comparison of biopsy samples with necropsy samples, there was complete regression of histopathological lesions in the intestine of one animal, which was culture negative at necropsy. Using a higher dose of MAP in a subsequent experiment resulted in a higher proportion of cases with histopathological lesions.

In conclusion, this is the first report of recovery from paratuberculosis in cattle. The experimental infection model provided realistic microbiological and histopathological outcome variation suitable for the study of the pathogenesis, treatment or prevention of Johne's disease.

1.4 PREDICTION OF THE GENETIC SUSCEPTIBILITY TO PARATUBERCULOSIS IN DAIRY CATTLE USING ALLELIC COMBINATIONS OF FIVE SINGLE NUCLEOTIDE POLYMORPHISMS (SNPS) IN CD209, SLC11A1, SP110, AND TLR2 GENES

María Canive[1], Rosa Casais[1], Patricia Vázquez[2], Ana Balseiro[3], José M. Prieto[3], Javier Amado[4], José A. Jiménez[5], Joseba Garrido[2], Ramón A. Juste[3], Marta Alonso-Hearn[2].

[1]*Servicio Regional de Investigación y Desarrollo Agroalimentario. Centro de Biotecnología, Deva, Asturias, Spain*
[2]*NEIKER- Instituto Vasco de Investigación y Desarrollo Agrario, Animal Health Department, Derio, Bizkaia, Spain*
[3]*SERIDA, Servicio Regional de Investigación y Desarrollo Agroalimentario. Centro de Biotecnología, Deva, Asturias, Spain*
[4]*LSAPA, Animal Heath Laboratory of the Principality of Asturias, Department of Microbiology, Gijón, Asturias, Spain*
[5]*CONAFE, Spanish Federation of Holstein Cattle, Madrid, Spain*

Recently, we have proposed that allelic combinations of five SNPs in four bovine genes (CD209, SLC11A1, SP110 and TLR2) could be associated with a low (LOWIN), latent (LATIN), or high (PATIN) risk of infection. Other possible combinations of these 5 SNPs were grouped in the average risk group (AVERIN).

The aim of this study was to validate the predictive ability of these multi-SNPs combinations and to estimate their associations with production traits in cows from a Spanish Holstein dairy farm (N= 99).

Of those 99 cows, 3 tested positive by fecal culture, 6 tested positive by ELISA and 9 tested positive by Map fecal PCR. Blood samples from all of the cows were collected for genomic DNA extraction, and the purified DNA was genotyped using the Illumina EURO G10K BeadChip.

Animals were then assigned to a LOWIN, LATIN, PATIN or AVERIN risk of infection and significant differences in the number of animals with a positive fecal culture were observed between the PATIN group (33 %) and the AVERIN (2.46 %), LATIN (0 %) or LOWIN (0 %) groups. In agreement with these results, significant differences in the mean log Map copies/gr of feces were observed between the PATIN and the other three groups. In addition, significant differences in mean ELISA ODs between PATIN (65.49) and AVERIN (15.97), LATIN (2.11), and LOWIN (3.27) groups were observed. Cows within the PATIN group had significantly higher milk fat yield, and superior udder, feet and legs and combined genetic scores than the other groups in two consecutive lactations. Superior genetic scores for production and morphological traits were also observed in PATIN cows (N=237, N=1409) within larger Spanish dairy populations (N=2776, N=15195) for two consecutive years.

The inclusion of paratuberculosis-associated markers in breeding programs should reduce culling and improve longevity, thus capturing greater potential milk yields from older cows.

1.5 EARLY DIAGNOSIS OF PARATUBERCULOSIS INFECTION USING ELISAS BASED ON THE DETECTION OF HOST BIOMARKERS

Rosa Casais[1], Crisitina Blanco[1], Ana Balseiro[1], Ramón Antonio Juste[1], José Miguel Prieto[1], Rosana Torremocha[2], Beatriz Soriano[3], Ricardo Ramos[2], Carlos Llorens[3], Javier Amado[4], Maria Canive[5], Marta Alonso[5]

[1]*SERIDA, Servicio Regional de Investigación y Desarrollo Agroalimentario. Centro de Biotecnología Animal, Deva, Asturias, Spain*
[2]*Parque Científico de Madrid, Unidad de Genómica, Campus de Cantoblanco, Madrid*
[3]*Biotechvana, Paterna, Valencia, Spain*
[4]*LSAPA, Laboratorio de Sanidad Animal del Principado de Asturias*
[5]*NEIKER- Instituto Vasco de Investigación y Desarrollo Agrario, Animal Health Department, Derio, Bizkaia, Spain*

Early detection of paratuberculosis (PTB) is essential to reduce disease transmission. Available diagnostic assays have poor sensitivities for the detection of early subclinical stages of PTB infection as *Mycobacterium avium* subsp. *paratuberculosis* (MAP) is excreted in low numbers and animals have low titers of specific antibodies. Host biomarkers have been considered as alternatives for early identification of the infection. Our recent work using transcriptomic analysis allowed the selection of 5 prognostic biomarkers (B1 to B5) overexpressed in blood and/or ileocecal valve of MAP infected Holstein cows with focal histopathological lesions in their gastrointestinal tissues

The aim of this study was to investigate the predictive ability of ELISAs designed for specific detection of bovine biomarkers (B1 to B5) to identify animals with latent forms of infection, characterized by the presence of focal histopathological lesions.

The diagnostic performance of these ELISAs was compared to that of the IDEXX ELISA for detecting MAP antibodies, and MAP specific fecal bacteriological culture and PCR. For the study, sera from 17 culled Holstein cows from a Spanish farm were used, 13 with focal lesions and 4 without histopathological lesions. The OD values obtained for each biomarker were subjected to ROC analysis. The sensitivity and specificity estimates for each ELISA were calculated based on the cut-off value which showed the highest diagnostic value (semisum of sensitivity and specificity

ELISA based on B1 detection had the highest diagnostic value (0.692) compared to those of the other ELISAs. B1-ELISA was able to detect cows with focal lesions with a higher sensitivity (38.46%) than the IDEXX ELISA (7.69%), the fecal bacteriological culture (0%) and the fecal PCR (18.18%), with the specificity of all these diagnosis techniques being 100%

In conclusion, measuring the concentration of the biomarker 1 in serum might be useful as a diagnostic tool for the early detection of MAP infection.

1.6 CHARACTERIZING RESPONSES OF IMMUNE CELL SUBSETS FROM M. PARATUBERCULOSIS (MAP) TEST POSITIVE AND TEST NEGATIVE COWS FROM COMMERCIAL HERDS TO MAP ANTIGEN STIMULATION IN VITRO.

Paul Coussens, Meredith C. Frie

Johne's disease (JD) has been particularly resistent to control and eradication efforts, which are hampered by a lack of approved vaccines and a poor understanding of protective immune responses. Another major obstacle to JD and MAP management is that JD is difficult to detect in many animals, in part due to variable immunity against MAP.

One of our goals is to improve knowledge of immune responses to MAP, identify possible correlates of protection, and to eventually relate these factors with genetic differences between cows.

In the present study, sample groups consisted of MAP test positive and MAP test negative cows from 8 commercial herds in Michigan. Peripheral blood mononuclear cells (PBMCs) from 154 MAP test negative and 96 age-matched MAP test positive cows were stimulated with MAP antigens in vitro. Following stimulation, subsets of CD4+, CD8+ and γδ T cells and B cells were examined using flow cytometry with CD25 (IL-2 receptor) expression as an activation marker.

In comparing MAP test positive to MAP test negative cows, MAP stimulation significantly increased CD25 expression on CD4+CD45R0+ T cells, on MHCII+ and MHCII- γδ T cells, and on SIgM+ B cells relative to cells from test negative cows. While clear and significant differences were detected between MAP test positive and MAP test negative cows following stimulation with MAP antigens, we were also able to detect a number of MAP test negative cows with subsets of PBMCs that responded to MAP antigens by upregulation of CD25. Differences between the T cell response of MAP test positive and negative cows were also muted due to cells from many test positive cows responding weakly or not at all to MAP antigens.

These results highlight the continum of peripheral immune responses to MAP in natural infections and difficulties in distinguishing JD positive and negative cows in commercial settings where prior exposure to MAP is likely.

1.7 COMPARISON OF SHEEP, GOATS, AND CALVES AS INFECTION MODELS FOR MYCOBACTERIUM AVIUM SUBSP. PARATUBERCULOSIS

Judith Stabel[1], John P. Bannantine[1], Jesse M. Hostetter[2].

[1]*USDA-ARS-National Animal Disease Center*
[2]*Iowa State University*

Animal infection models to study *Mycobacterium avium* subsp. *paratuberculosis* (MAP) infection are useful for evaluating the efficacy of vaccines and other therapeutics for the prevention or treatment of infection. Because Johne's disease is highly prevalent in US dairy cattle populations much of our research utilizes neonatal calves as infection models. Disadvantages to a calf model are the expense and that they do not generally progress to clinical disease within a 12-month period.

The goal of the present study was to compare sheep, goats, and calves as infection models, perhaps demonstrating clinical signs within a shorter time span.

In the present study neonatal sheep, goats, and calves (n=4) received 109 cfu of a cattle isolate of MAP in milk replacer on days 0, 3 and 6 of a 12-month study and sampled monthly during the study period.

Results demonstrated a robust antigen-specific IFN-g response at 90 days post-inoculation for sheep and goats with lower responses noted for calves. By 360 days, IFN-g responses were 50 and 82% higher for calves than for goats and sheep, respectively. Although MAP-specific antibody responses were first observed in sheep at 90 days, calves had higher antibody responses throughout the remainder of the study. Following pass-through shedding on day 7, fecal shedding was fairly negligible across treatments but remained higher for calves throughout the study. Colonization of tissues was variable within treatment group but trended higher for calves and sheep for the majority of tissues. Interestingly, upon stimulation of PBMCs with a whole-cell sonicate of MAP, flow cytometric analyses demonstrated greater populations of CD4+ T cells and lower populations of NK and CD14+ cells for goats and calves compared to sheep.

Although sheep and goats have many similar properties as infection models to calves, overall calves proved to be a more standardized model with appropriate fecal shedding, tissue colonization and host immune responses upon long-term infection.

1.8 LDLR FACILITATES MYCOBACTERIAL INFECTION-INDUCED CHANGES TO HOST LIPID METABOLISM.

Matt D. Johansen[1], Auriol C. Purdie[1], Kumudika de Silva[1], Karren M. Plain[1], Stefan H. Oehlers[2].

[1]*University of Sydney*
[2]*Centenary Institute*

Host lipids play an important role in the establishment of mycobacterial infection. Pathogenic mycobacteria such as M. tuberculosis and M. leprae are capable of sequestering and utilising cholesterol within host macrophages, representing an important intracellular survival mechanism. Previously, we have demonstrated that macrophages infected with MAP are able to accumulate host cholesterol and modulate specific host lipid metabolism pathways. Of particular interest, is the modulation of key host lipid metabolism gene, Low Density Lipoprotein Receptor (ldlr), during early MAP infection. This gene plays an important role in the uptake and processing of low density lipoproteins (LDL) within macrophages and is thought to facilitate foam cell formation in mycobacterial infection. However, targeted gene depletion studies are not viable in ruminant models of MAP infection.

In this study, we have utilised the natural host-pathogen pairing of the zebrafish-*M. marinum* platform to investigate the functional significance of ldlr in the establishment of mycobacterial infection.

Gene depletion of ldlr was achieved with morpholino microinjection in zebrafish embryos at the 1-4 cell stage. At approximately 30 hours post-fertilisation, embryos were infected with Mycobacterium marinum via caudal vein injection with bacterial burden and lipid abundance measured at 3 and 5 days post-infection.

Following depletion of ldlr, there was a significant reduction in the bacterial burden at both 3 and 5 days post-infection. Analysis of the total lipid content demonstrated that ldlr-deficient embryos had a markedly decreased concentration of LDL compared to infected controls. Examination of the lipid density within granulomas identified significantly decreased foam cell abundance in ldlr-deficient embryos.

The findings from this study demonstrate that ldlr plays a primary role in facilitating mycobacterial infection via macrophage transdifferentiation to foam cells.

1.9 EFFECTS OF VACCINATION ROUTE ON THE IMMUNE RESPONSE IN THE RABBIT PARATUBERCULOSIS INFECTION MODEL

Iraia Ladero[1], Rakel Arrazuria[1], Elena Molina[1], Miguel Fernández[2], Marcos Royo[2], Joseba Garrido[1], Ramón Juste[1], Valentín Pérez[2], Natalia Elguezabal[1]

[1]*NEIKER-Instituto Vasco de Investigación y Desarrollo Agrario, Animal Health Department, Derio, Bizkaia, Spain*
[2]*Departamento de Sanidad Animal, Instituto de Ganadería de Montaña (CSIC-Universidad de León), Facultad de Veterinaria, C/ Profesor Pedro Cármenes s/n, 24071 León*

Commercially available paratuberculosis (PTB) vaccines are applied subcutaneously, but have some drawbacks regarding interference with diagnostic tests. This problem could be circumvented if protection could be afforded by more specific immunization.

To explore the effect of more local or reduced antigen administration, oral and intradermal vaccination routes were evaluated in the rabbit PTB infection model.

Tweenty five NZW rabbits were divided into experimental groups of five: 2 non-vaccinated controls, challenged and non-challenged; and 3 vaccinated groups treated with commercial subcutaneous (VS) or with experimental oral (VO) or intradermic (VI) vaccines. Oral challenge with Map K10 was performed one month after vaccination. Peripheral humoral response (PPA-3 ELISA) was analyzed throughout the experiment. At 5 months post-infection, rabbits were killed and submitted to histopathological scoring and T cell response assessment by splenocyte lymphoproliferation assay. Severity of granulomatous lesions focusing on gut associated lymphoid tissue was evaluated and immunohistochemistry was performed on sacculus rotundus and vermiform appendix sections in order to phenotypically characterize macrophages (calprotectin, CD163 and TNF-alpha) and to evaluate the local expression of IFN- gamma.

Higher PPA-3 antibody titers and lymphoproliferative indexes were detected in the VS group compared to the VO and VI groups. Highest lesion scores were recorded in the challenged non-vaccinated group. Lesion score reduction was 94.11%, 19.41% and 58.82% for VS, VO and VI respectively. Both VO and VS presented highest levels of CD163. Calprotectin and IFN-gamma correlated with lesion scores in all vaccinated groups being highest in the VO groups followed by VI and VS ($p<0.001$).

Both alternative routes showed some degree of histopathological protection, and although lagging behind that provided by the commercial subcutaneous vaccine, this proved the concept that immunization can be induced by them. Fine tuning of the new routes is worth in order to match the effects of dose and adjuvants in the commercial vaccine.

1.10 CYTOKINE GENE EXPRESSION PROFILES IN THE BLOOD OF CLINICAL AND NON-CLINICAL SHEEP AND GOATS INFECTED WITH MYCOBACTERIUM AVIUM SUBSPECIES PARATUBERCULOSIS.

Ganesh Gangaram Sonawane[1], Bhupendra Nath Tripathi[1], L.L.L. Prince[2], Rajendra Kumar Meena[1].

[1]*ICAR-National Research Centre on Equines, Hisar-125001(Haryana), India*
[2]*ICAR - Directorate of Poultry Research, Rajendranagar, Hyderabad 500 030 (Telangana), India*

Paratuberculosis is a chronic, infectious disease of domestic and wild ruminants characterized by granulomatous enteritis, progressive emaciation and irreversible diarrhea. It is caused by *Mycobacterium avium* subspecies *paratuberculosis* (MAP) a hardy, slow growing acid-fast microorganism. The role of cytokines in pathogenesis of paratuberculosis, in sheep and goats has been rarely studied in India.

The present study was aimed to evaluate expression profiles of certain cytokine genes in the blood of MAP infected and non-infected sheep and goats maintained in the intensive system of management.

A total of 385 adult sheep and 55 adult goats of the paratuberculosis endemic and non-endemic farms respectively, were screened for MAP infection using multiple diagnostic tests. On the basis of clinical symptoms, ELISA, PCR and faecal examination, these animals were divided in to different groups as MAP infected clinical form (ICF, 9.8% sheep), MAP infected non-clinical form (INCF, 5.9% sheep) and MAP negative (NI, 84.3% sheep). Amongst 55 adult goats 6 were (4.5%) in the clinical form (ICF) of the disease while remaining 49 goats (95.5%) were negative (NI) for MAP. The healthy sheep from non-endemic farm, for paratuberculosis served as control animals (CS) for the study. These sheep were negative for MAP infection by multiple tests. Blood samples were collected from representative animals from each group and relative gene expression levels of TGF-β, IL-10, IFN-γ and TRAF-1 cytokines were evaluated by using qPCR.

In sheep, the comparison between various groups revealed significantly higher mean relative expression of TGF-β, IL-10 and TRAF-1 in the NI sheep as compared to the ICF, INCF and CS groups. The mean relative expression level of IFN-γ in NI and INCF sheep was found to be comparable; however, it was significantly higher than the ICF and CS sheep. In the ICF sheep, the expression levels of all the cytokines under study were found to be down regulated with significant decrease in IL-10 level. The comparison of various cytokines between NI and ICF goats did not exhibit significant difference in the mean relative expression levels of all the cytokines determined in the study.

From the present study, it can be concluded that the immunological response in the MAP negative sheep was stronger than MAP infected non-clinical and clinical sheep of the paratuberculosis endemic farm. The trend of cytokine expressions due to MAP infection in sheep and goats in the present study indicate different immunological response in both the species and required further comprehensive studies including identification of resistance markers to map infection.

1.11 CONTROL OF JOHNE'S DISEASE IN A NEW ZEALAND DAIRY HERD USING TEST AND CULLING.

Frank Griffin[1], Rory O´Brien[1], Simon Liggett[1], Andrew Bates[2].

[1]Disease Research Limited, Invermay Agricultural Centre, Mosgiel, New Zealand
[2]Vetlife, Centre for Dairy Excellence, Geraldine, New Zealand

Serial serum ELISA and faecal PCR (fPCR) tests were carried out in a large NZ dairy herd (1,200 cows) suffering high losses (3-5%) from clinical Johne's Disease. In years 2013, 2014, 2015 & 2016 whole herd serum ELISA was carried out on all cows in the herd. All cows with ELISA (+) test results had fPCR tests carried out and all cows shedding >103 MAP genomes/Gm faeces) were culled before calving.

After whole herd ELISA testing and selective culling of fPCR (+) animals the prevalence of ELISA (+) reactivity dropped each year; 26.5%, 10.1%, 6.8% to 3.3%. The prevalence of fPCR (+) animals among ELISA (+) animals ranged from 7-21%. Clinical deaths decreased from 80 cows pa in the year testing began, to

Milk production increased significantly after the first year of testing and culling, where production levels (KgMilk Solids) increased from 452 to 490KgMS per cow. There was no difference in Days in Milk between E(+) and E(-) animals. Milk solids in Kgs per cow per day (1.53) was significantly lower (P

This case study involved a large New Zealand dairy herd suffering significant capital costs from clinical Johne's disease (>$NZ 100K pa). Whole herd serological ELISA testing identified 'at-risk' animals and serial fPCR identified shedders This allowed stratification of the 'high-risk' shedder animals for culling. Selective culling of shedder animals resulted in control of infectious spread and a dramatic increase in animal health and production within one year.

This case study involved a large New Zealand dairy herd suffering significant capital costs from clinical Johne's disease (>$NZ 100K pa). Whole herd serological ELISA testing identified 'at-risk' animals and serial fPCR identified shedders This allowed stratification of the 'high-risk' shedder animals for culling. Selective culling of shedder animals resulted in control of infectious spread and a dramatic increase in animal health and production within one year.

1.12 RECOMBINANT SECRETARY PROTEINS BASED 'COCKTAIL ELISA' TO DIFFERENTIATE BETWEEN INFECTED AND VACCINATED BOVINES AGAINST MYCOBACTERIUM AVIUM SUBSPECIES PARATUBERCULOSIS

Kundan Kumar Chaubey[1], Shoor Vir Singh[1], Rinkoo Devi Gupta[2], Kirti Dua[3], Shukriti Sharma[3], Niranjana Sahoo[4].

[1]*Animal Health Division, Central Institute for Research on Goats, Makhdoom, PO- Farah, Mathura, UP, India*
[2]*Department of Life Sciences and Biotechnology, South Asian University, New Delhi, India*
[3]*Department of Epidemiology and Preventive Medicine, College of Veterinary Science and Animal Husbandry, GADVASU, Ludhiana, Punjab, India*
[4]*Department of Epidemiology and Preventive Medicine, College of Veterinary Science and Animal Husbandry, OUAT, Bhubaneswar, Odisha, India*

In India Johne's disease is endemic in the domestic livestock population and bio-load of *Mycobacterium avium* subsp. *paratuberculosis* in domestic livestock has shown increasing trend. For the control of Johne's disease using 'Indigenous Vaccine', a test to differentiate between infected and vaccinated animals (DIVA) is mandatory.

Study estimated potential of six recombinant secretary proteins based 'cocktail ELISA' as 'marker DIVA assay'. Fecal and serum of 178 cows from 6 Gaushalas were collected (2013–2016). On the basis of health and two tests (fecal microscopy and Indigenous ELISA) cows were taken as negative (healthy-112) and positive (infected-66) for Johne's disease. Of 112 healthy cows, 44 were vaccinated (Indigenous Vaccine) and rest 68 were negative control. Cocktail ELISA using of six recombinant secretary proteins (MAP 1693c, MAP 2168c, MAP Mod D, MAP 85c, MAP Pep AN and MAP Pep AC) was compared with 'Indigenous ELISA'. OD values were expressed as sample-to-positive ratio. Sensitivity and specificity of ELISA tests were determined. Mc Nemar's test and kappa agreement were applied to estimate statistical significance of the results.

Testing of 178 cows by microscopy and i_ELISA, 66 and 112 were positive and negative, respectively at zero day. At conclusion of study highest percent (61.7%) of cows were positive by i_ELISA, followed by microscopy (37.0%) and c_ELISA (35.3%). C_ELISA was validated as DIVA test, by screening of 178 well categorized cows (Infected-66, healthy-68 and vaccinated-44) were screened by two ELISA tests. Positive cows (44) in c_ELISA were partitioned as negatives in i_ELISA. It was due to the fact that 'Indigenous Vaccine' uses only 'In-activated MAP' bacilli, so it will not elicit expression of antibodies against 'secretary antigens'. I_ELISA and c_ELISA had sensitivity of 100.0 and 59.1% and specificity of 57.2 and 100.0%, respectively. P value by Mc-Nemar test was 0.0001 and strength of agreement was moderate using Kappa value at confidence interval of 0.401–0.611.

'Cocktail ELISA' using 6 recombinant secretory proteins successfully differentiated infected, vaccinated and healthy cows.

1.13 IMMUNE RESPONSE INDUCED BY A LOCAL STRAIN OF MYCOBACTERIUM AVIUM SUBSP. PARATUBERCULOSIS IN ANIMAL MODELS.

María Alejandra Colombatti Olivieri[1], Roberto Damian Moyano[1], María José Gravisaco[1], Gabriel Eduardo Traveria[2], María de la Paz Santangelo[1], María Isabel Romano[1].

[1]*Instituto de Biotecnología, INTA Castelar, Buenos Aires, Argentina*
[2]*Centro de Diagnostico e Investigaciones Veterinarias (CEDIVE) de la Facultad de Ciencias Veterinarias, Universidad de La Plata, Buenos Aires, Argentina*

A commercial vaccine against paratuberculosis, like Silirum, reduces clinical disease and fecal shedding of *Mycobacterium avium* subsp. *paratuberculosis* (Map) but do not prevent the infection.

The aim of this study is the evaluation of the immune response and protection induced by a local strain administered as an inactivated vaccine in mouse and immune response in cattle.

Mouse model: 3 groups of 15 female BALB/c mice A) unvaccinated group, B) Silirum vaccine, C) inactivated local strain with incomplete Freund's adjuvant. Two vaccine doses of 0.2 mL (2.5 mg dry bacteria pellet /mL) were administered subcutaneously with a difference of 15 days. One month after the last vaccine dose, mice were intraperitoneally challenged with 1x108 CFU of avirulent Map strain. At pre-challenge cytokines and IgG were measured, and at 6 and 12 weeks post-challenge the animals were sacrificed and the spleen was taken for CFU counting and liver for histopathology. Bovine model: 3 groups of 5 calves (1-2 months of age): 1) unvaccinated group, 2) Silirum vaccine 3) inactivated local strain with Montanide (Seppic SA) adjuvant. Serum samples were taken monthly for the measurement of antibodies and every 2 months whole blood was extracted to measure cytokines and lymphocyte populations. Also the intradermal reaction with PPDa and PPDb were evaluated.

In the mice model, the groups vaccinated with the local strain showed CFU counts and score lesion lower than the control group. The immune response was predominantly humoral (IgG1) and cellular response was also detected, with production of IFNg, IL-6 and IL-2, in both vaccinated groups. In cattle the immune response was higher (IgG and IFNg) for the group vaccinated with the local strain and there were no reactors to PPDb in the intradermal reaction.

The results indicate that our local strain protected mice from challenge with a virulent strain inducing a humoral and cellular immune response, and this response was higher than the commercial vaccine in the bovine model.

1.14 MAP-SPECIFIC VOLATILE ORGANIC COMPOUND PROFILE: A COMPARATIVE ANALYSIS OF THREE DIFFERENT IN VITRO STUDIES.

Anne Küntzel[1], Sina Fischer[1], Andreas Bergman[2], Peter Oertel[2], Philip Trefz[2], Wolfram Miekisch[2], Jochen Schubert[2], Petra Reinhold[1], Heike Köhler[1].

[1]*Friedrich-Loeffler-Institut, Institute of Molecular Pathogenesis, Germany*
[2]*University of Rostock, Department of Anesthesia and Intensive Care, Germany*

Diagnosis of *Mycobacterium avium* subsp. *paratuberculosis* infection is currently done by cultural isolation followed by identification via polymerase chain reaction. Thus, the established diagnostic protocols comprise several steps and therefore are time-consuming, labor-intensive and expensive. Detection of volatile organic compounds (VOCs) above bacterial cultures is considered to be a promising approach to accelerate cultural identification.

The aim of this project was to define a MAP-specific VOC profile by comparing results of three different in vitro studies which were focused on biological and methodological variability of MAP related VOCs.

In all three studies VOCs were measured above different MAP strains, all cultivated on Herrold's Egg Yolk Medium and above non- inoculated slants of pure medium. For pre-concentration of VOCs, needle-trap micro-extraction was employed. Samples were subsequently analyzed using gas chromatography-mass spectrometry. All volatiles were identified and calibrated by analyzing pure reference substances.

More than 100 VOCs were detected in the headspace above inoculated and control slants in each study. Each time, about 30–40 substances were indicating bacterial growth of MAP. Twelve of these volatiles were found to be specific for MAP in all three studies. Further 18 substances were included in the MAP-specific VOC profile in two of three studies.

This Data supports the hypothesis that MAP emits a variety of metabolites in different concentration levels, which may lead to MAP- specific biomarkers.

1.15 EX VIVO VACCINATION WITH A PLGA/MPLA NANOPARTICLE VECTORED A 35 KDA MAJOR MEMBRANE PROTEIN FROM MYCOBACTERIUM AVIUM PARATUBERCULOSIS ELICITS KILLING OF INTRACELLULAR BACTERIA BY CD8 T CELLS

Cleverson D Souza[1], Gaber S. Abdellrazeq[1], Mahammoud M. Elnaggar[2], John P. Bannantine[3], Julianne Hwang[4], William C. Davis[2].

[1]*Washington State University, Faculty Veterinary Medicine, Alexandria University, Egypt*
[2]*Vet Micro/Pathol, Washington State University, Pullman, WA, U.S., Faculty Veterinary Medicine, Alexandria University, Egypt*
[3]*National Animal Disease Center, USDA-Agricultural Research Service, Ames, US*
[4]*Veterinary Clinical Sciences, Washington State University, Pullman, US*

Analysis of the immune response to a *Mycobacterium avium paratuberculosis* (Map) candidate relA deletion mutant vaccine in cattle revealed a 35 kD membrane protein (MMP) is a major component of the immune response to Map.

Studies were conducted to determine the functional activity of effector T cells that develop following stimulation with blood dendritic cells (bDC) and monocyte derived DC (MoDC) pulsed with MMP alone or incorporated into nanoparticles (NP) comprised of poly lactic-co-glycolic acid and monophosphoryl lipid A (PLGA/MPLA).

Effector T cells were generated ex vivo from monocyte depleted PBMC (mdPBMC) by 2 rounds of stimulation, first with bDC then with MoDC pulsed with MMP alone or incorporated into PLGA/MPLA NP. Primed mdPBMC were incubated with monocyte derived macrophages (MoMΦ) uninfected and infected with Map, then processed to determine the level of killing mediated by cytotoxic lymphocytes (CTL). Flow cytometry was used to phenotype primed PBMC and MoMΦ.

Comparison of mdPBMC stimulated with MMP alone or incorporated in PLGA/MPLA NP showed the CTL proliferative response was enhanced when mdPBMC were primed with MPLA vectored MMP. Depletion studies demonstrated CD8 CTL activity only developed if CD4 and CD8 T cells were present at the time of Ag presentation. Intracellular killing was mediated through the perforin granzyme granulysin pathway. Development of CD8 CTL is complex and involves coordinate stimulation of CD4 and CD8 T cells by antigen presenting cells.

The ex vivo CTL response to MMP is enhanced when incorporated into a nanoparticle vector.

1.16 DISEASE STATE INFLUENCES THE PRESENCE OF MACROPHAGES AND MYCOBACTERIUM AVIUM SUBSP. PARATUBERCULOSIS IN BOVINE INTESTINAL TISSUE.

Caitlin Jenvey[1], Jesse M. Hostetter[1], Adrienne L. Shircliff[2], John P. Bannantine[2], Judith R. Stabel[2].

[1]*Iowa State University*
[2]*USDA-ARS-National Animal Disease Center*

Johne's disease is an enteric disease caused by the intracellular pathogen *Mycobacterium avium* subsp. *paratuberculosis* (MAP). Upon ingestion of MAP, it is translocated across the intestinal epithelium and taken up by intestinal macrophages. Once ingested, MAP may be killed by macrophages, or depending upon the bacterial burden and immunological status of the animal, MAP may thwart innate defense mechanisms and persist within the macrophage.

This study aimed to correlate the presence of macrophages and MAP in bovine mid-ileal tissue with stage of infection.

Immunofluorescent (IF) labeling was performed on frozen bovine mid-ileal intestinal tissue collected from 28 Holstein dairy cows. Macrophages were labeled using a monoclonal anti-macrophage surface antigen (AM-3K) and MAP was labeled using a polyclonal rabbit heat-killed MAP antigen, with IF labeling visualized using a confocal microscope. Imaging software was used to quantify the surface area and intensity of IF labeling.

The presence of macrophages within the mid-ileal tissue sections was higher for clinical cows, followed by subclinical cows and then uninfected control cows. Macrophages were present throughout the intestinal tissue in clinical cows, including the inner muscle layer, submucosa, crypt and villi ends, while presence of macrophages in both subclinical and control cows were limited to the submucosa and inner muscle layer. Clinical cows also demonstrated significantly higher MAP SA and macrophage and MAP co-localization SA, when compared to subclinical cows, and was present within the submucosa and crypt lamina propria, progressing into the villi ends in some clinical cows.

Our findings indicate that number of macrophages increases with progression of disease, however, a significant number of the macrophages present in the mid-ileum are not associated with MAP. This suggests that although the bovine innate immune system is sufficiently stimulated to recruit macrophages in response to MAP invasion, the macrophages of clinically infected cows are ultimately unable to clear MAP, resulting in disease progression and clinical presentation.

1.17 MACROPHAGE ANALYSIS TO IDENTIFY FUNCTIONAL GENETIC MARKERS ASSOCIATED WITH JOHNE'S DISEASE.

Nathalie Bissonnette[1], Jean-Simon Brouard[1], Olivier Ariel[1]Eveline Ibeagha-Awemu[1], Filippo Miglior[2].

[1]*Agriculture and Agri-Food Canada, Sherbrooke Research and Development Centre, Sherbrooke, QC, Canada*
[2]*Centre for Genetic Improvement of Livestock, University of Guelph, Canadian Dairy Network, Guelph, ON, Canada*

Johne's disease (JD) is a debilitating chronic disease in ruminants caused by *Mycobacterium avium* ssp. *paratuberculosis* (MAP), which manipulates gut macrophages as a strategy for its survival and dissemination. The mechanisms of JD pathogenesis are still unclear, but it is known that MAP subverts the host immune system by using macrophages as its primary reservoir.

A genomics study used primary macrophages from 22 infectious (MAP-shedding) cows and 28 JD-negative cows for the identification of functional genetic variants.

The transcriptome of their blood-circulating primary macrophages was analyzed using next-generation RNA sequencing (RNA-Seq). Genetic variations in those macrophages from susceptible cows were studied in an in vitro MAP infection assay. DNA genotypes were also identified using a BovineSNP50 beadchip, imputed to a high-density (HD) DNA chip. Functional variant effect was predicted using SnpEff.

A total of 1.36 million high-quality variants were identified: 60% were identified by RNA-seq, but only 3.6% were common to the DNAchip, making both technologies highly complementary. Of the 814,168 RNA-seq variants derived from primary macrophages, 93,279 were indels. Interestingly, 2,435 RNA-seq variants are predicted to produce high functional effects on known genes, but only 33 DNA-HD genotypes were found in this category. A genome-wide association study identified two expression quantitative trait loci (eQTLs) on BTA4 and 11 genomes at $-\log10\ (P) \geq 7$. Network and pathway analysis revealed interesting cue regarding pathways that deserved further investigation (e.g. STAT transcription factor).

In this study, we successfully identified eQTLs and the regulatory pathways that distinguish between MAP-infected and JD-negative cows. Integration of our findings (genomic information) into conventional programs for young sire selection and progeny testing could yield better, more accurate and rapid genetic improvements for resistance to bovine paratuberculosis in Canadian dairy herds.

Poster

1.1 FACTORS AFFECTING MAP SAMPLE-TO-POSITIVE RATIO IN THE MILK OF IRISH DAIRY COWS

Conor Mc Aloon, Lorna Citer.

A significant effect of season, in addition to other factors such as milk yield, days in milk (DIM) and somatic cell count (SCC) may impact the optimal thresholds used to determine an animal as positive. In addition, such information may inform optimal sampling times to most accurately define the infection status of an animal.

Therefore, the aim of this study was to investigate factors affecting the milk sample to positive ratio (S/P) at cow level by examining the association between test month, milk yield, DIM and SCC.

Data were extracted from the Animal Health Ireland pilot programme consisting of 42,657 milk recordings from 18,569 cows across 187 dairy herds over the 24-month period from January 2014 to December 2015 inclusive. The effect of month of the year, milk yield, DIM and lnSCC was investigated using a generalised linear mixed model with restricted cubic splines to account for non-linear variables. S/P ratio for each test was natural log transformed for normality. Random effects were used to account for within animal and within herd clustering and separate models were constructed for primi- and pluriparous animals.

In the heifer model, analysis demonstrated significantly decreased S/P values in April, May and June in 2014 but not 2015. There was no effect of milk yield on S/P value. However, serological response peaked at early and late lactation with a nadir between 100 and 200 DIM. S/P increased in a curvilinear manner with increasing lnSCC. In the pluriparous model, significant decreases in S/P response were observed in April, May and June again in 2014 but not in 2015. In addition, S/P decreased linearly with increasing milk yield and increased in a curvilinear manner with increased lnSCC. The S/P value also peaked at the beginning and end of lactation with lowest values from 20 to 200 DIM.

These findings represent important considerations in the interpretation of the serological value in Irish dairy herds.

1.2 IMMUNO-PROTEOMIC ANALYSIS OF SECRETORY PROTEINS OF NOVEL 'INDIAN BISON TYPE' BIOTYPE (STRAIN 'S 5') OF MYCOBACTERIUM PARATUBERCULOSIS AND ITS DIAGNOSTIC SIGNIFICANCE IN DOMESTIC LIVESTOCK ENDEMIC FOR JOHNE'S DISEASE

Saurabh Gupta[1], Shoor Vir Singh[1], A.K. Bhatia[2], Shivalingappa Yamanappa Mukartal[3], Doddamane Rathnamma[3], Kratika[1]

[1]*Animal Health Division, Central Institute for Research on Goats, Makhdoom, PO- Farah, Mathura, UP, India*
[2]*Department of Biotechnology, GLA University, Mathura, India*
[3]*Department of Veterinary Microbiology, KVAFSU, Bengaluru, India*

Johne's disease caused by *Mycobacterium avium* subspecies *paratuberculosis* (MAP) has adverse impact on the livestock productivity and is responsible for huge economic losses to the livestock farmers world-wide, in terms of reduced milk and meat production, early culling and persistence of infection in animal herds. Due to chronic nature and occurrence of four stages of disease from silent to advance clinical stages, its diagnosis is still a major challenge. Diagnostic efficacy of imported commercial ELISA kits containing antigen(s) of MAP not allied to our strain, is uncertain. In India, this is first report using naturally secreted proteins of indigenous strain ('S 5') for the development of ELISA for detection of MAP in native domestic livestock population.

To study the immuno-proteomic profile and to estimate their diagnostic potential of natural secretory proteins of novel 'Indian Bison Type' biotype (strain 'S 5') of MAP.

Analysis of harvested secretory proteins was done by SDS gel electrophoresis followed by 1D and 2D Immuno-blotting.

Immuno-proteomic analysis showed that six secretary proteins (28, 31, 34, 38, 45 and 56 kDa) were immunogenic and consistently recognized by polyclonal rabbit anti-sera at 4-12 weeks of initial growth period. Four secretory proteins (apparent molecular mass 17, 19, 47 and 65 kDa) appeared only in early growth period of MAP (4 and 6 weeks). Combined application of these proteins was evaluated by 'Indirect ELISA' in search of potential antigenic epitopes. Secretory proteins of 4 and 6 weeks were used as 'antigen candidate' and 8.5 to 13.5% & 17.3 to 19.1% increase in specificity was recorded when compared with whole cell sonicated semi-purified protoplasmic antigen (sPPA) using caprine and bovine serum samples, respectively.

Indigenously developed assays using secretory protein(s) of native strain ('S 5') may be used both as 'antigen and vaccine candidate' and serve as the backbone of future control programs in the country. Using imported commercial ELISA kits, we in India are unknowingly under reporting the incidence of MAP infection in domestic livestock.

1.3 DEEP TRANSCRIPTOMIC PROFILING OF PRIMARY BOVINE MACROPHAGES IN COWS WITH JOHNE'S DISEASE SUGGESTS A CHANGE IN HOST LIPID METABOLISM AND A STATE OF INNATE IMMUNE TOLERANCE

Nathalie Bissonnette[1], Olivier Ariel[1]Nicolas Gévry[2], Gilles Fecteau[3], Eveline M. Ibeagha-Awemu[1]

[1]*Agriculture and Agri-Food Canada, Sherbrooke Research and Development Centre, Sherbrooke, Qc, Canada*
[2]*Université de Sherbrooke, Sherbrooke, Qc, Canada*
[3]*Department of Biology, Faculty of Veterinary Medicine, Université de Montréal, Saint-Hyacinthe, Qc, Canada*

Mycobacterium avium ssp. *paratuberculosis* (MAP) is the causative agent of Johne's disease (JD), which is also known as paratuberculosis, in ruminants. The mechanisms of JD pathogenesis are still unclear, but it is known that MAP subverts the host immune system by using macrophages as its primary reservoir.

In our study, ex vivo experiment using primary monocyte-derived macrophages (MDM) allowed a deep analysis through next- generation sequencing (RNA seq) of the transcriptome of primary macrophages and in response to live MAP infection.

The gene expression signatures of primary macrophages from cows diagnosed as negative or positive for JD [JD (–) or JD(+), respectively] revealed differential host–pathogen interplay, highlighting long-term mechanisms established during mycobacterial disease. In JD (–) cows, a considerable number of genes (1,436) were differentially expressed at 8 h post-infection (hpi). Interestingly, while expected immune response pathways were significally challenged by MAP, others related to hepatic fibrosis/hepatic stellate cell activation, lipid homeostasis, such LXR/RXR (liver X receptor/retinoid X receptor) activation, and autoimmune diseases such rheumatoid arthritis were consistently among the top significant pathways. Unexpectedly, the macrophages from JD (+) cows did not present a clear response pattern to MAP infection, neither during the early period (1–8 hpi) nor at 24 hpi. Analysis of the transcriptomic profile of the uninfected macrophages revealed 868 genes that were differentially expressed in JD (+) versus JD(–) cows. The down-regulated genes predominantly modified the general cell metabolism by down-regulating amino acid synthesis, cholesterol biosynthesis, and other energy production pathways while introducing a pro-fibrotic pattern associated with foam cells. These modifications show the apparent importance of the metabolic pathways hijacked by MAP to subvert the immune cells.

Our findings support the hypothesis that MAP could induce a tolerant-like state in circulating monocytes when they are differentiating into macrophages. Thereby, MAP could further promote disease persistence by influencing macrophage behavior in the long term. This report contributes to a better understanding of MAP control of immune cells and its mechanisms of survival.

1.4 DESIGN AND OPTIMIZATION OF A POLYPROTEIN FOR DIAGNOSIS OF BOVINE PARATUBERCULOSIS

Maria Isabel Romano[1], Natalia Alonso [1], Moyano Roberto Damian[1], Maria Laura Mon[1], Natanael Griffa[1], Magali Romero[2], Fiorela Alvarado Pinedo[2], Maria de la Paz Santangelo[1], Gabriel Traveria[2]

[1]*INTA-CONICET*
[2]*CEDIVE*

Paratuberculosis control requires identification and removal of infected animals from the herd. To emphasize the PTB controls on health systems it is necessary to develop new solutions to diagnosis. Currently, the gold standard technique is bacterial culture; however, it is a complex strategy given the slow growth of these mycobacteria

In this context, the aim of the present work is to develop a diagnostic tool based on an ELISA and a lateral flow immunochromatography (LFIC) that allows the specific diagnosis of paratuberculosis.

For that purpose, antigens from Mycobacterium avium subspecie paratuberculosis previously identified and characterized by our group were designed and synthesized as polyprotein tagged with His. The expression of the construction was performed in E coli BL21 and subsequently purified using a nickel column. For the development of the ELISA, 100ul of the polyprotein (0,25mg/ml) was used and 98 sera were evaluated corresponding to healthy animals, paratuberculosis-infected cattle and tuberculosis -infected cattle. In parallel, the same sera were analyzed with another ELISA frequently used in the laboratory for the diagnosis of paratuberculosis using PPA-3 as antigen.

Our findings showed that while ELISA-PPA-3 reacts with sera from animals infected with M. bovis, while the polyprotein was found to be sensitive and specific for diagnosis of bovine paratuberculosis. For the development of the LFIC, G protein was conjugated to colloidal gold, purified bovine immunoglobulins (0, 05µg/µl) was used as control line and the polyprotein (0, 5 µg/µl) was used as test line. The same sera were evaluated and a correlation was achieved with the results of the ELISA using the polyprotein

Finally, the results obtained so far show that the constructed polyprotein can be used for the specific diagnosis of paratuberculosis since it did not induce reactions with sera from tuberculosis-infected cattle or from healthy cattle, one of the main problems of the methods used nowadays.

1.5 EVALUATION OF SEROLOGICAL TESTS FOR THE DETECTION OF PARATUBERCULOSIS IN ITALIAN BUFFALOS (BUBALUS BUBALIS): A CLASS LATENT APPROACH

Matteo Ricchi[1], Giorgio Galletti[1], Simone Russo[2], Fabrizio Gamberale[3], Esterina DeCarlo[4], Norma Arrigoni[2]

[1]*Istituto Zooprofilattico Sperimentale della Lombardia e dell'Emilia Romagna, Reparto di Sorveglianza Epidemiologica dell'Emilia-Romagna (SEER), Bologna Italy.*
[2]*Istituto Zooprofilattico Sperimentale della Lombardia e dell'Emilia Romagna, National Reference Centre for paratuberculosis, Podenzano (PC), Strada Faggiola 1, Italy.*
[3]*Istituto Zooprofilattico Sperimentale del Lazio e della Toscana 'M. Aleandri', Via Appia Nuova 1411, 00178 Roma, Italy.*
[4]*Istituto Zooprofilattico Sperimentale del Mezzogiorno, National Reference Centre for Hygiene and technologies of water buffaloes farming and production, Italy*

The ELISA test is the most used assay for paratuberculosis control and it is utilised in the Paratuberculosis Italian Guidelines for assigning the health ranking to cow and buffalo herds. Currently, available commercial ELISA kits for the diagnosis of paratuberculosis in buffaloes are not supported by robust validation data. The gold standard recommended by the OIE Terrestrial Manual for the in vivo diagnosis of paratuberculosis is the cultural assay. However, the sensitivity of these tests is low, not suitable to be used as standard in indirect test validation. Some researchers have proposed PCR tests as gold-standard because of its sensitivity and rapidity.

Aim of this work was the validation of commercial ELISA tests aimed at detecting paratuberculosis infected buffaloes. The project was designed to collect data from herds located in various areas with different prevalences.

So far, the sampling was carried out in Frosinone and Rome provinces from two different herds, one with a low and another with a higher prevalence. The blood and faeces of 449 buffaloes were analysed in parallel by Id-Vet ELISA test and IS900-qPCR. In order to evaluate the accuracy/performance of both tests, these preliminary data were analysed by a Bayesian two latent class model, combining different cut off for both tests. The model included strong prior only for performance of qPCR (Se 50%, 90% sure is between 40% and 60%; Sp 70%, 95% sure is higher than 50%).

Results showed that at each combination of cut off considered (ELISA 0.6 vs PCR 38, 36, 34 Cq; ELISA 0.7 vs PCR 38 Cq) Sp of both tests was very high, with Posterior Median always higher than 95% and narrow high density intervals (HDI). Conversely, the model failed in evaluating Se: for ELISA, the HDI are too wide, for PCR, the strong prior was not modified and, finally, the prevalences estimated by the model are similar, in contrast with what requested by the model.

We believe further samplings will improve the model.

1.6 MAMMARY EPITHELIAL CELLS PREVIOUSLY INFECTED BY ESCHERICHIA COLI ALTERS THE MYCOBACTERIUM AVIUM SUBSP. PARATUBERCULOSIS TRANSLOCATION INFECTION

David Germano Gonçalves Schwarz[1], Fernanda Miyagaki Shoyama[1], Leandro Licursi Oliveira[2], Srinand Sreevatsan[1], Maria Aparecida Scatamburlo Moreira[2]

[1]*Michigan State University*
[2]*Universidade Federal de Viçosa*

Mastitis and paratuberculosis are diseases that cause great economic losses in the world's dairy industry. It is known that Mycobacterium avium subsp. paratuberculosis (MAP) can be released by the milk intermittently. However, it is still unclear whether the E. coli mastitis could alter the permeability of mammary epithelial cells increasing their evasion to the alveolus of the mammary gland.

The aim of this study was to verify the dynamic between MAP and E. coli in the cellular response of bovine mammary epithelial cells (MAC-T).

Transwell assay was used to analyze MAP translocation in a previous E. coli infection in MAC-T cultivated at 37ºC in a humidified chamber. On the upper chamber, E. coli was inoculated by 30 min, washed and then MAP was added in the lower chamber for 10, 30 and 120 min. The supernatant of the upper chamber was collected and analyzed by absolute real-time PCR and culture (MB7H9). From MAC-T cells, RNA was extracted and qPCR was conducted on three different targets: IL-10, MAPK p38 IL-1β and β-actin was used as a housekeeping gene. Ratio between target/reference was calculated on a Roche Lightcycler 480II.

Our experiments showed that in MAC-T cells infected by E. coli, MAP is rapidly attracted from baso to apical site of the cells up 30 min of exposure, decreasing at 120 min. No colonies characteristics of MAP in MB7H9 medium were observed and no significant expression of IL-10 and MAPK p38 was detected. However, IL-1β was upregulated at 120 min in cells previously infected by E. coli.

Our findings suggest that E. coli presence in the mammary epithelial cells could attract MAP from distant sites and facilitate its release into apical side of mammary epithelial cells. Besides that, the subsequent upregulation of IL-1 β could attract MAP to inside the cells at 120 min. This is the first report of signaling contribution in MAP dynamics in MAC-T cells in with a mastitis agent.

1.7 PROTOTYPE OF THE REFERENCE STANDARD FOR THE QUANTIFICATION OF MYCOBACTERIUM AVIUM SUBSP. PARATUBERCULOSIS BY QUANTITATIVE PCR IN FAECES

Petr Kralik[1], Monika Beinhauerova[1], Iva Slana[1]

[1]*Veterinary Research Institute*

Quantitative PCR (qPCR) is nowadays a frequently employed method for the direct detection and quantification of Mycobacterium avium subsp. paratuberculosis (MAP). However, the quantity of MAP assessed by qPCR is dependent on the type of the qPCR quantification standard and the way how DNA quantity of the standard is determined. The most common way is the absorbance determination, which can, however, differ among different types of DNA standards (plasmid, isolated DNA). This leads to the situation that quantities of MAP determined by the different qPCR assays in different laboratories are not comparable with each other.

The objective of this study was to prepare the prototype of the reference standard for the quantification of MAP in faeces by qPCR. This standard would make possible to compare the results between individual laboratories across the world.

Supernatant of the 40% faceal suspension was artificially contaminated with MAP reference strain, mixed with the matrix and lyophilized. Different types of matrices were tested. Repeatability and stability of the MAP reference standard in time were tested by qPCR.

This first generation of the MAP reference standard was prepared to assess the suitability of selected procedure for the generation of the worldwide qPCR standard. Our laboratory is ready to serve as the supplier of the MAP reference standard for diagnostic laboratories.

This effort should bring the unification of the DNA quantification standards (independently on the DNA isolation and qPCR) in qPCR detection and quantification of MAP. And should lead to the situation that it will be possible to directly compare quantitative data from different qPCR assays and different laboratories. This work was supported by the Ministry of Interior of the Czech Republic (VI20152020044) and the Ministry of Agriculture of the Czech Republic (RO0518).

1.8 SENSITIVITY OF AURAMINE-RHODAMINE AND ZIEHL-NEELSEN STAINING FOR MYCOBACTERIUM AVIUM SUBSP. PARATUBERCULOSIS DETECTION IN LIQUID CULTURE FROM ENVIRONMENTAL FECAL SAMPLES.

Enrico Inama[1], Marta Mainenti[1,] Paola Fiorin[1], Annalucia Tondo[1], Marta Vascellari[1], Nicola Pozzato[1]

[1]Istituto Zooprofilattico Sperimentale delle Venezie

Paratuberculosis is a chronic disease featured by granulomatous enteric disease, weight loss, chronic diarrhea and death. The reference standard for the diagnosis of Mycobacterium avium subsp. Paratuberculosis (MAP) is culture in solid medium. Liquid culture systems are extensively used for their shorter turnaround time but the detection systems lack in sensitivity and specificity and serial testing by PCR is required. Preliminary acid-fast staining is usually performed but the sensitivity of these tests is unknown.

To evaluate the sensitivity of the Ziehl-Neelsen (ZN) staining method and Truant TB-Auramine-Rhodamine (AR) fluorescent staining method in non-automated liquid cultures.

Among 517 environmental fecal samples examined by direct Real-Time PCR and culture in enriched Middlebrook 7H9 medium (7h9+) followed by confirmatory Real-Time PCR, we selected a group of 56 samples that tested negative at the first molecular test, and positive with a Ct value >25.0 at the second molecular analysis. From each 7H9+ sample, we prepared and stained two slides of smears by ZN and AR. Both sets of slides were double blinded and examined in parallel by two readers.

Acid-fast bacilli were identified by either one or both operators in 49/56 samples stained with AR (87.50% sensitivity), and in 48/56 stained with ZN (85.71% sensitivity). These results show that AR and ZN stainings were useful to detect MAP in 7H9+ containing 104-105 MAP/ml of 7H9+ broth that represent 101-102 MAP/g of feces.

Despite both microscopic methods were not able to detect all the Real-Time positive samples, the AR fluorescent staining is easier to use than Ziehl-Neelsen staining also for less experienced personnel. Their use within the diagnosis process is under evaluation.

1.9 DISEASE PROGRESSION IN SUSCEPTIBLE VERSUS RESILIENT MYCOBACTERIUM AVIUM SUBSPECIES PARATUBERCULOSIS (MAP) INFECTED COWS

Nathalie Bissonnette[1], D. Kelton[2], G. Fecteau[3], P. Griebel[4], F. Miglior[5]

[1]*University of Guelph, Guelph, ON, Canada*
[2]*Ontario veterinary college*
[3]*Faculté de médecine vétérinaire, Université de Montréal, Saint-Hyacinthe, QC, Canada*
[4]*School of public health, University of Saskatchewan, Saskatoon, SK, Canada*
[5]*Centre for Genetic Improvement of Livestock, University of Guelph, Canadian Dairy Network, Guelph, ON, Canada*

Not all cow exposed to MAP progress to clinical Johne's disease (JD). MAP can subvert host immune responses and host genetic variation is expected to contribute to differences in immune responses following pathogen exposure.

A longitudinal study was performed to monitor disease progression in MAP infected cows. Twenty-three high prevalence herds (10 free stall and 13 tie-stall) were selected from among 92 farms in Ontario and Quebec, Canada.

Farms were selected using high environmental MAP load as a criteria. Cows over 2 years old were followed for a 3 to 5 year period. Blood and fecal samples were collected 2-3 times/year and tested for MAP-specific antibodies (MAP-Ab; IDEXX) and direct qPCR (ISMAP02), respectively. MAP in feces was also detected by culture (MGIT Para TB). Body condition score and disease information were recorded.

Among a total of 3,537 cows, 2,259 cows received a diagnosis retrospectively on the basis of MAP test results. With a total of 509 JD cows, herd prevalence ranged from 5.6% to 24.3%. Cows classified diseased were excreting 5×103-5 CFU/g feces during the last two consecutive sampling periods. Herd mean MAP-Ab values, calculated using the peak value from each infected cow, varied significantly among herds but not among geographic regions. A second group of 263 exposed cows (>6 years old) was identified with significant MAP-Ab titres but an absence of detectable MAP in feces throughout the entire 3-5 year period. A third group of 222 cows was identified as intermittent MAP shedders but they did not progress to clinical JD while remaining lactating proficient cows.

Classification of subclinical MAP infections is challenging. The present longitudinal study provides evidence, however, for distinct phenotypes of MAP susceptibility or resilience when using the combined criteria of serum antibody and fecal shedding. This phenotypic classification is relevant when performing genomic analyses to identify markers associated with susceptibility/resilience to JD.

1.10 CYTOTOXIC T CELLS WITH ABILITY TO KILL INTRACELLULAR BACTERIA ARE ELICITED BY ANTIGEN PRESENTING CELLS PULSED WITH A MEMBRANE PROTEIN FROM MYCOBACTERIUM AVIUM PARATUBERCULOSIS

Cleverson D Souza[1], Gaber S Abdellrazeq[1], Mahmoud M Elnaggar[1], John, P Bannantine[2], William C Davis[3],

[1]*Washington State University, Pullman, WA, U.S, Faculty Veterinary Medicine, Alexandria University, Egypt*
[2]*National Animal Disease Center, USDA- Agricultural Research Service, Ames, IA, U.S*
[3]*Vet Micro/Pathol, Washington State University, Pullman, WA, U.S.*

Analysis of the immune response to a Mycobacterium avium paratuberculosis (Map) deletion mutant, ΔrelA, revealed deletion abrogated the capacity of the mutant to establish a persistent infection. Studies conducted with ΔrelA revealed vaccination elicited cytotoxic T cells (CTL) with the ability to kill intracellular bacteria.

As described herein, further studies were conducted ex vivo to determine whether the immunogenic target of the response to ΔrelA was directed towards a 35 kDa major membrane protein, MMP.

Flow cytometry was used to phenotype T cell subsets proliferating in response to antigens presented by APC pulsed with ΔrelA or MMP. A single gene probe, F57, specific for Map was used with qPCR and Propidium monoazide (PMA) to monitor CTL activity of responding cells. Equivalent CTL responses were obtained with APC primed with ΔrlA and with MMP. Depletion studies demonstrated CD8 T cells were the main T cell subset with CTL activity. Further studies demonstrated CTL activity only developed if there was coordinate Ag presentation to CD4 and CD8 T cells by APC pulsed with MMP. Intracellular killing was mediated through the perforin, granzyme, granulysin pathway.

Flow cytometry was used to phenotype T cell subsets proliferating in response to antigens presented by APC pulsed with ΔrelA or MMP. A single gene probe, F57, specific for Map was used with qPCR and Propidium monoazide (PMA) to monitor CTL activity of responding cells. Equivalent CTL responses were obtained with APC primed with ΔrlA and with MMP. Depletion studies demonstrated CD8 T cells were the main T cell subset with CTL activity. Further studies demonstrated CTL activity only developed if there was coordinate Ag presentation to CD4 and CD8 T cells by APC pulsed with MMP. Intracellular killing was mediated through the perforin, granzyme, granulysin pathway.

Results indicate that MMP is a good candidate for development of a vaccine against Map.

1.11 MYCOBACTERIUM COMPLEX IN INDIAN LIVESTOCK: DEVELOPING DIAGNOSTICS FOR RESOURCE LIMITED AREAS

Rathnagiri Polavarapu[1], Rudrama Devi Punati[1], Prudhvi Chand Mallepaddi[1], Revathi Poonati[1], Mukta Jain[2], Jagdip Singh Sohal[2]

[1]*Genomix Molecular Diagnostics Pvt. Ltd., Hyderabad, India|*
[2]*Amity Center for Mycobacterial Disease Research, Amity University Rajasthan, Jaipur*

Mycobacterial infections like bovine tuberculosis (BTB) and paratuberculosis are major infectious diseases of ruminant species. Both zoonosis and reverse zoonosis have been reported for mycobacterial infections. In India both diseases have been found to be endemic in livestock and burden is continuously increasing. Existing diagnostic methods for mycobacteria are very difficult to implement in field level and it requires well-established and highly expensive equipment with expert technical professionals. Conventional culture methods are less sensitive, time taking and PCR has poor sensitivity. Recent studies indicate that anti- mycobacterial antibody response can be used to detect infected animals in early as well as latent phases. Serological assays utilizing cocktail of antigen should effectively cover the diversity of immune responses and provide more accurate tools for immunologic diagnosis of Mycobacterium complex. Present study focused on to facilitate the early diagnosis of mycobacterial infections using Point of Care diagnostics such as Lateral flow assay (LFA) and ELISA.

To develop lateral flow assay (LFA) and plate ELISA based diagnostic kits for simultaneous diagnosis of bovine tuberculosis and paratuberculosis in livestock

We used specific recombinant secretory proteins from Mycobacterium tuberculosis complex & M. bovis (ESAT6, CFP10, MBP70 and MBP83) and five specific proteins (CF041, CF040, CF341, CF281 and MAP2168) from M. avium ssp. paratuberculosis, M. phlei PPD (nonpathogenic mycobacteria) and M. bovis PPD (pathogenic mycobacteria) to develop LFA and ELISA based sero diagnostics for simultaneous diagnosis of tuberculosis and paratuberculosis. The diagnostic performance of these kits was validated using serum samples from 3264 animals (cattle- 2354, buffaloes- 201, sheep- 684 and camel- 25).

The overall sensitivity is 81.78% and specificity is 90.12% for Lateral flow assay, and for ELISA the sensitivity is 88.98% and a specificity of 95.99% were obtained.

Present tests LFA and ELISA can differentiate paratuberculosis and tuberculosis infection. Developed lateral flow and ELISA methods using cocktail of recombinant proteins are very simple, rapid, inexpensive, sensitive and specific diagnostic assays for detecting the Mycobacterium complex infection in animals thus fulfills the requirements of resource limited settings.

1.12 EVALUATION OF FOUR COMMERCIAL REAL-TIME PCR SYSTEMS FOR DETECTION OF MYCOBACTERIUM AVIUM SUBSP. PARATUBERCULOSIS FROM BOVINE FAECAL SAMPLES

Giovanni Ghielmetti[1], Roger Stephan[1], Natasha Carroli[2], Ute Friedel[2], Simone Scherrer[2], Sarah Schmitt[2]

[1]*Institute for Food Safety and Hygiene, University of Zurich, Switzerland*
[2]*Institute of Veterinary Bacteriology, University of Zurich, Switzerland*

Early stage detection of Mycobacterium avium subsp. paratuberculosis (Map) infection is fundamental to accelerate disease eradication programs. Faecal culture remains the most sensitive technique according to the literature, but it is not suitable for the rapid identification of animals shedding the pathogen and for screening of whole herds.

Different diagnostic methods such as real-time qPCR systems targeting various genes have been proposed as valid alternatives to culture. Most commercial qPCR systems target the IS900 insertion elements, although highly homologous insertion elements have been described in environmental Mycobacterium species. The diagnostic accuracy of commercial kits was evaluated on routine specimens originating from naturally infected and healthy animals.

Four DNA extraction kits combined with qPCR systems targeting IS900 provided by four different manufacturers (Qiagen, IDEXX, IDvet and Gerbion) were used for the analysis of 15 bovine faecal samples. Each specimen has been tested in duplicate, investigated for the presence of the f57-gene using an in-house qPCR and analysed on solid and liquid culture media. In addition, in order to assess the specificity of the different qPCR systems, a panel of 30 mycobacterial stains including species reported to display homologous IS900 insertion elements were tested.

In comparison with the qPCR targeting the f57-gene the following analytical Se and Sp were determined for the four commercial kits: 0.9 – 1.00 Qiagen, 1.00 – 0.66 IDEXX and 1.00 – 0.50 for IDvet and Gerbion. Despite the small number of samples analysed and taking into account that none of the 30 tested non-Map strains lead to a false-positive result, overall Se and Sp differed remarkably between the four kits.

The current study demonstrates the importance of choosing the appropriate methodology for the detection of Map through bovine faecal samples and its impact on diagnostic accuracy.

1.13 EFFECT OF FEEDING HEAT-TREATED COLOSTRUM ON CELL-MEDIATED IMMUNE RESPONSE TO MYCOBACTERIUM AVIUM SUBSP. PARATUBERCULOSIS IN NEONATAL DAIRY CALVES

Fernanda Miyagaki Shoyama[1], Juliana Lie Taya[1], Scott J Wells[2], Judy R Stabel[3], Maxim C Cheeran[2], Srinand Sreevatsan[4], Sandra M Godden[2]

[1]*Faculdade de Medicina do ABC, Sao Paulo - Brazil*
[2]*University of Minnesota - USA*
[3]*National Animal Disease Center, U.S. Department of Agriculture, Ames, IA – USA*
[4]*Michigan State University - USA*

Dairy calves can acquire Mycobacterium avium subsp. paratuberculosis (MAP) infection through the consumption of contaminated milk and colostrum. Heat treating (HT) colostrum (60°C-60 min) may be one management strategy to reduce risk of MAP infection via colostrum.

Test whether dairy calves fed HT colostrum at birth would have a reduced likelihood of a positive CMI test, suggesting a lower exposure/infection rate to MAP, as compared to calves fed fresh (FR) colostrum.

A convenience sample of five Midwest Holstein dairy herds with confirmed Johne's disease were selected. 250 newborn calves were enrolled and randomized to be fed either FR or HT colostrum. Blood samples were collected at 4weeks and 4months of age. Cells were stimulated with whole-cell sonicated MAP antigen, Concanavalin A (positive control) and PBS (negative control). Bovigam® ELISA assay was used to measure IFN-γ levels. The results were expressed as optical density of T-cell responses and categorized as positive or negative for MAP exposure after normalization with PBS. FR and HT colostrum samples were tested for MAP on culture and by direct PCR (IS900). Logistic regression models, accounting for random effect of calf within herd, were used to investigate the association between CMI test results (Pos/Neg) at 4wks and 4mos of age and the following risk factors: a)fed HT or FR colostrum, b)fed MAP-neg or MAP-pos colostrum.

CMI results were available for 223 calves (HT=111, FR = 112). The proportion of calves fed MAP-pos colostrum was significantly lower for calves fed HT (7.2%) versus FR (37.5%) colostrum (P=0.008). The odds of a neg CMI test did not differ between calves fed HT or FR colostrum at 4wks (HT=87.4%; FR=92.0%; OR=0.58, 95%CI:0.16, 2.08, Pvalue=0.3) or at 4mos (HT=91.9%; FR=94.6%; OR=0.68, 95%CI:0.13,3.58, Pvalue=0.55). The odds of a neg CMI test did not differ among calves fed MAP-Pos vs MAP-Neg colostrum at 4wks (MAP Pos=94.0%; MAP Neg=88.4%; Pvalue=0.35) or at 4mos (MAP Pos=96.0%; MAP Neg=92.5%; Pvalue=0.73).

While feeding HT colostrum decreased the risk of MAP exposure to calves, treatment had no effect on CMI-positivity in dairy calves at 4wks and 4mos of age. Similarly, feeding MAP-Pos (vs MAP-Neg) colostrum had no association with CMI-positivity at both ages. These results raise the question of whether or not the CMI test is an appropriate test to indicate MAP exposure/infection in calves of this age due to the immaturity of the calf on immunological responses. Furthermore, the low number of IFN-γ positive calves (4wks=24; 4mos=15) suggests that calf exposure was low, regardless of colostrum treatment.

1.14 TRANSCRIPTIONAL PROFILING OF ILEOCECAL VALVE AND WHOLE BLOOD OF HOLSTEIN DAIRY COWS AT DIFFERENT STAGES OF MYCOBACTERIUM AVIUM SUBSP. PARATUBERCULOSIS (MAP) INFECTION

Marta Alonso-Hearn [1], Maria Canive[1], Rosana Torremocha[2], Ana Balseiro[3], Beatriz Soriano[4], Ricardo Ramos[2], Carlos Llorens[4], Rosa Casais[3]

[1]*NEIKER-Instituto Vasco de Investigación y Desarrollo Agrario, Animal Health Department, Derio, Bizkaia, Spain*
[2]*Scientific Park of Madrid, Genomic Unit, Campus de Cantoblanco, Madrid, Spain*
[3]*SERIDA, Servicio Regional de Investigación y Desarrollo Agroalimentario, Centro de Biotecnología, Deva, Asturias, Spain*
[4]*Biotechvana, Paterna, Valencia, Spain*

Biomarkers that can accurately define disease progression for Map are still to be identified.

In this study we examine gene expression in ileoceal valve (ICV) and whole blood of Holstein cows that were naturally infected with Map. The animals were at different stages of disease to identify transcriptional changes and cellular pathways that may be critical to the progression from focal histopathological lesions to a more advanced diffuse paucibacillary and diffuse intermediate categories (n=5 for each group).

All control animals were negative for Map by ELISA, PCR and culture. Approximately 250 ng of total RNA was used for RNA-Seq libraries creation using the NEBNext® Ultra Directional RNA Library preparation kit (Illumina). cDNA libraries were sequenced using Illumina NextSeq technology. Trimmed reads were aligned to the Bos Taurus genome with TopHat and Cufflinks was used to identify differentially expressed (DE) genes.

The most DE genes were identified in ICV of animals with diffuse intermediate lesions and in blood of animals with focal lesions. Many of the cellular pathways that had strong differential gene expression in ICV between uninfected and infected cows were related to the innate immune and inflammatory responses. These pathways only had differential expression in blood samples from animals with focal lesions. It is important to note that in each of the group comparisons they were more upregulated genes compared to down- regulated genes in ICV. This is contrary to the comparisons with blood samples where more genes were downregulated. Eight putative biomarkers were lesion-specific and upregulated in both blood and ICV of cows with focal lesions, seven in cows with paucibacillary lesions and five in cows with intermediate lesions.

Our study provides insights into how cattle respond to a natural Map infection within two infection targets. The identified genes can be studied further to aid in better diagnostic tools and vaccines.

1.15 COST-EFFECTIVE PARATUBERCULOSIS MANAGEMENT WITH AN EASY-TO-USE QPCR ON POOLS OF FAECES TO REDUCE RE-INFECTION PRESSURE

Laffont Mathieu[1], Grewis Lise[1], Forero Rafael[1], Greatrex Anna[1], Pourquier Philippe[1]

[1]*IDvet Genetics*

IDvet Genetics has developed an easy-to-use Map detection system on faeces, including a rapid sample preparation kit (EZPREP), and a qPCR test, ID GeneTM Paratuberculosis Duplex (IDMAP).

The IDMAP qPCR test may be used for relative quantitative testing, for individual or pooled testing. It allows the detection of chronic shedders as well as intermittent or passive carriers, and may be used as a decision-making tool to manage positive animals.

Three independent herds were tested. Herd 1 (n=161) had a sero-prevalence of 10%, herd 2 (n=127) of 10% and herd 3 (n=76) of 14%. All animals were tested individually using the ID Screen® Paratuberculosis Indirect ELISA for sera. Faeces samples were first tested individually, then pooled by 5 and 10 using EZPREP and IDMAP qPCR.

For herd 1, positive individuals were pooled with 4 or 9 negative samples. Here the use of pooling by 5 or 10 identifies animals with an individual Cq inferior to 33 which would not be detected by serological testing. For herds 2 and 3, samples were randomly pooled by 5 or 10. Results demonstrate that animals may be detected positive by pooled PCR when at least one chronic or intermittent shedder (Cq<33) is present inside the pool.

In order to minimize the cost of analysis while maintaining a detection of chronic and intermittent shedders or passive carriers, the use of pooled samples for a qPCR analysis may be a relevant way of analysis. The method may be used in different contexts depending on the herd status and aim of the control program. It permits to help the farmer to make a decision to cull the animal or not, depending on the need to lower the Map pressure in the herd. Nevertheless with a seroprevalence <2%, it is recommended to test individual animals to eliminate the last shedders.

Keywords: PCR, pooling, high shedders.

1.16 SEROPREVALENCE OF PARATUBERCULOSIS IN GOAT AND BOVINE CATTLE OF THE STATE OF SAN LUIS POTOSI, MEXICO

Sosa-Martínez Luz Elena[1], Lara-García Luis[1], Vega-Manriquez Xochil[1], Morón-Cedillo Felipe[1], Castrellón-Ahumada Victoria[2], Maldonado-Castro Edith[2], Herrera- Corredor Alejandra[1], Ballesteros- RodeaGilberto[1], Santos-Díaz Rosa Elena[1], Chávez-Gris Gilberto[2]

[1]*Facultad de Agronomía y Veterinaria, Universidad Autónoma de San Luis Potosí*
[2]*Facultad de Medicina Veterinaria y Zootecnia, Centro de Enseñanza Investigación y Producción Animal en Altiplano, UNAM*

Paratuberculosis, also known as Johne's disease, is a pathology caused by Mycobacterium avium subsp. paratuberculosis (MAP), which is characterized by granulomatous enteritis, its presence in livestock production units represents significant economic losses for the livestock sector. Understanding the epidemiology of Johne's disease is important to determining the management and control strategies of the disease in susceptible species of productive livestock

The aim of the present study was to determinate the seroprevalence of MAP in goats and bovine cattle from the Potosino Highlands, through the ELISA-P35 serological test

Goat herds and bovine cattle herds from different municipalities of the state were selected. Ten percent of the population of the herd were taken, obtaining 131 samples of goat blood serum and 159 of bovine cattle.

Subsequently, the ELISA P35 test was performed to determine the prevalence, showing 8.32% in goats, with very similar results in the different herds tested (Cerritos of 8.20%, Matehuala 8.43% and Villa de Hidalgo of 8.18%). Similar results were obtained for bovine cattle in the municipalities of Villa de Arriaga of 0.74%, and Soledad de Graciano Sánchez of 0.53%, while in San Nicolás Tolentino the prevalence was 3.15%, obtaining a prevalence in cattle of 1.42%.

Our findings are the first results that provide relevat information to determinate the prevalence of paratuberculosis in the State of San Luis Potosí and propose the implementation of sanitary strategies that allow to control the disease in the livestock productions.

1.17 SEROPREVALENCE OF PARATUBERCULOSIS IN GOAT HERDS OF THE STATE OF SAN LUIS POTOSÍ, MÉXICO.

Gilberto Ballesteros Rodea[1], Luis Alberto Lara García[1], Delia Xóchil vega Manriquez[1], Gilberto Chávez Gris[2]

[1]Facultad de Agronomía y Veterinaria, Universidad Autónoma de San Luis Potosí
[2]Facultad de Medicina Veterinaria y Zootecnia, Centro de Enseñanza Investigación y Producción Animal en Altiplano, UNAM

Paratuberculosis, also known as Johne's disease, is a pathology caused by the bacterium Mycobacterium avium subsp. paratuberculosis (MAP), which is characterized by a chronic intestinal lesion of proliferative character. Paratuberculosis was first described in Germany by Johne and Frothingham in bovines in 1895, in this species it has been well characterized and widely distributed in most countries, especially in dairy herds. MAP, is an intracellular facultative pathogen, acid-alcohol resistant and dependent on mycobactin for its growth in vitro. Johne's disease is a severe granulomatous gastroenteritis, with lymphangiectasis and associated lymphangitis, the final consequence is the appearance of a typical syndrome of malnutrition, with chronic and progressive weight loss, with the presence of chronic or intermittent diarrhea, affects ruminants, mainly cattle, sheep and goats, producing a chronic and incurable enteritis that irremediably ends with the life of the animal. The infection is caused by ingestion of colostrum, milk or fecal matter of bacteria-eliminating animals. The infected animals go through a long subclinical stage, during which, small amounts of MAP are eliminated by milk and feces intermittently. Its identification is complicated, mainly commercial packages that evaluate the production of gamma interferon by lymphocytes in culture are used.

Determination of seroprevalence in goat herds of the State of San Luis Potosí, Mexico.

The serum of goats from three different regions of San Luis Potosi, Mexico was obtained to perform the diagnosis of paratuberculosis through the ELISA diagnostic test, using the P35 antigen as capture antigen.

In the goat herds of the three different regions of San Luis Potosí, a prevalence of 8.32% of paratuberculosis was obtained, using the ELISA diagnostic test with the P35 antigen.

This work is the first exploratory report in the diagnosis of paratuberculosis in goats in the State of San Luis Potosí, Mexico. The development of a health program based on the needs of each herd is recommended.

Parcial Supported by UNAM, DGAPA PAPIIT IT201118

1.18 PREVIOUS CONTACT OF GOAT KIDS WITH CORYNEBACTERIUM PSEUDOTUBERCULOSIS OR MYCOBACTERIUM AVIUM SUBSP HOMINISUIS INFLUENCES THE OUTCOME OF MYCOBACTERIUM AVIUM SUBSP PARATUBERCULOSIS INFECTION

Marcos Royo[1], Miguel Fernández[1], Iker A Sevilla[2], Noive Arteche[1], Raquel Vallejo[1], Miguel Fuertes[1], Daniel Gutiérrez-Expósito[1], M. Carmen Ferreras[1], Julio Benavides[1], Natalia Elguezabal[2], Valentín Pérez[1]

[1]*Universidad de León*
[2]*NEIKER*

Pathogenesis of paratuberculosis can be influenced by different factors such as host response, strain of Mycobacterium avium subsp paratuberculosis (Map) that causes infection, or previous contacts with Map or other bacteria by the host.

The study investigates the possible role that previous oral sensitization with antigenically Map-related bacteria plays on the pathogenesis of caprine paratuberculosis in an experimental model.

Twenty-three, 1.5 month-old goat kids were orally inoculated with 5x103 cfu of Mycobacterium avium subsp hominisuis (Mah) or Corynebacterium pseudotuberculosis (Cptb). Two months after, five kids from each treated group, and another 5 non-inoculated, were orally challenged with 1.8x1010 cfu of Map. Peripheral cellular (IFN-? release assay) and humoral immune responses were analysed. At 5 months post-infection, pathological examination was carried out. Severity and location of Map-related granulomatous lesions were evaluated by means of granuloma counting in the intestine and related lymph nodes.

Specific paratuberculosis antibody or cell-mediated peripheral responses were detected only in the Map-infected kids, with no differences between them. Granulomatous lesions were detected only in the three Map-infected groups. While in kids infected only with Map lesions were seen at more locations (including jejunum with no lymphoid tissue), in animals previously infected with Mah or Cptb, lesions were restricted to the lymphoid tissue. However, the highest number of granulomas were seen in the Cptb+Mah group and the lowest in kids from the Mah+Map group.

In the conditions of the study, previous oral infection with low doses of Mah or Cptb has an influence on the development of lesions at early states of Map infection in goats. Although with differences in intensity, while in only Map-infected kids lesions progress to areas of intestinal mucosa devoid of Peyer's patches, in the rest of the groups they tend to be restricted to the lymphoid tissue.

1.19 STUDY ON THE INTERFERENCE OF PARATUBERCULOSIS VACCINATION ON THE TUBERCULOSIS ERADICATION PROGRAM IN GOATS IN CASTILLA Y LEÓN (SPAIN)

Miguel Fernández[1], Ana Grau[1], Raquel Vallejo[2], Noive Arteche[2], Marcos Royo[2], Daniel Gutiérrez-Expósito[2], M. Carmen Ferreras[2], Julio Benavides[2], Olga Mínguez[1], Valentín Pérez[2]

[1]*Junta de Castilla y León*
[2]*Universidad de León*

Vaccination is a well recognized method for paratuberculosis control. However, it has been pointed out that it can affect the results of the tuberculosis control programs based on the use of single intradermal skin test, due to the cross reaction that vaccinated animals can show.

This study evaluates the results of the tuberculosis eradication program during one year, in relation with the vaccination status of the goat herds.

Single intradermal tuberculin (SIT) test, using bovine PPD, has been performed in a total of 1744 goat herds in Castilla y León (Spain) during 2017, according to the official procedures. Vaccination status was assessed considering the herd history and records. Bacteriological culture of the lung and respiratory lymph nodes, for Mycobacterium caprae isolation was carried out in the positive animals. Moreover, the ileocaecal valve and regional lymph nodes were examined by histopathological methods for paratuberculosis lesions detection.

From the 1744 herds examined, 184 (10.5%) had been vaccinated against paratuberculosis. SIT test positive reactors were detected in 34 goat herds (1.9% of the total). From them, 15 herds were vaccinated (8.1% of the vaccinated herds). In particular, only two out of the 42 herds vaccinated and tested within 6 months after vaccination were positive to the SIT test. From the 15 vaccinated herds showing SIT test reactors, M. caprae was isolated in 6 of them (40%). Regarding paratuberculosis lesions, they were present in a 30.8% of vaccinated animals and were limited to the lymphoid tissue; however, in non-vaccinated herds they were detected in a 61.8% and diffuse lesions were seen in at least four goats.

According to these results, interference of paratuberculosis vaccination on the tuberculosis control programs, by the use of SIT test, only would occur in a low number of herds and decrease with the time.

1.20 STANDARDIZATION OF ELISA-P35 FOR DIAGNOSIS PARATUBERCULOSIS FROM GOAT'S MILK

Alejandra Montserrat Hernández-Guerra[1], Victoria Elizabeth Castrellón-Ahumada[1], Abel Manuel Trujillo- García[1], Isabel Estévez Denaives[1], Gilberto Chávez-Gris[1]

[1]*CEIEPAA, Facultad de Medicina Veterinaria y Zootecnia de la Universidad Nacional Autónoma de México. Km 8,5 carretera Tequisquiapan-Ezequiel Montez*

Paratuberculosis is a disease of worldwide distribution and produce a negative economic impact. The standardization of an ELISA- P35 test, complement the diagnosis of paratuberculosis detecting positive animals as individual test, but also the response to Map could be detected in bulk tank.

Develop and standardize ELISA-P35 for the diagnosis of paratuberculosis from goat milk.

ELISA-P35 in milk was standardized using as antigen P35 protein (10 µg/ml). The sensitivity, specificity values were calculated with respect to IS900 PCR. Samples of milk, blood and feces were taken from 143 goats. Bacteriological culture was performed from feces. DNA extraction was performed from milk and blood using commercial silica membrane columns. The detection of mycobacteria in samples of blood, feces and milk was performed using the IS900 PCR test. The detection of antibodies against Map from ELISA-P35 in plasma samples was performed.

In this study 66.43% were positive by ELISA-P35 in plasma, while 53.84% were positive by ELISA-P35 in milk. Of 143 animals, only two were positive for fecal culture, which were detected as positive in milk and plasma ELISA, which agrees with reports that the level of serum antibodies is directly related to the probability that the animal eliminates in the feces the bacterium and is detected by the fecal culture. Two animals were positive by culture, were also positive both in ELISA in milk and in serum, but negative in the PCR tests of feces, milk and blood. These results could be due to DNA extraction protocols or intermittent elimination of Map in blood and milk. Map was also identified directly by PCR IS900 in 6 milk samples and 5 blood samples. Of 11 positive animals by PCR IS900, 63.6% were positive in the ELISA in milk and 45.4% were positive in the ELISA in serum, suggesting that P35 ELISA-P35 in milk is able to detect a greater number of infected animals when they are producing milk. Animals considered positive in PCR and fecal culture, 9 were positive to ELISA-P35 in milk and of 130 animals were considered negative, 68 were positive to ELISA-P35 in milk obtaining 69.23% and 47.69% of sensitivity and specificity respectively.

The ELISA in milk is a useful method to detect individuals that in the future could develop the disease. A positive result in the ELISA, either in serum or in milk, must be confirmed using culture or PCR.

This study was supported by the project UNAM, DGAPA PAPIIT IT201118

1.21 COMPARISON OF AN ANTIGEN DERIVED FROM MYCOBACTERIUM AVIUM WITH THE PROTOPLASMAL ANTIGEN OF MYCOBACTERIUM AVIUM SUBS. PARATUBERCULOSIS (PPA-3) FOR THE DIAGNOSIS OF PARATUBERCULOSIS THROUGH ELISA.

Edith Maldonado Castro[1], Verónica Blásquez Vázquez[2], Gabriel Campos Montes[3], Carolina García Sánchez[1], Carolina Segundo Zaragoza[1], Gilberto Chávez Gris[1]

[1]*CEIEPAA; FMVZ, Universidad Nacional Autónoma de México.*
[2]*FMVZ, Universidad Nacional Autónoma de México.*
[3]*Universidad Autónoma Metropolitana.*

Paratuberculosis is caused by *Mycobacterium avium* subsp. *paratuberculosis* (Map) and affects ruminants causing a granulomatous enteritis, has also been associated with Crohn's disease in humans. Map belongs to the Mycobacterium avium Complex where *Mycobacterium avium* subsp. *avium* (Maa) is also found. Hurley et al., in 1989, described a genetic identity between the different subspecies of the Mycobacterium avium complex (95-100%), likewise Bannantine et al. found that Map's genomic sequence shows 97% identity with respect to Maa. Due to the high degree of homology described, Maa proteins have been used in ELISA, as an alternative in the diagnosis of paratuberculosis. In Mexico and other countries, PPA-3 has been used in ELISA, however, currently a problem at the national level is the need to use other antigens as an alternative in serological diagnosis, due to the difficulty and import costs.

The objective was to develop an indirect ELISA testing different concentrations of Maa for the identification of seropositive animals and compare it with the PPA-3 antigen in the serodiagnosis of paratuberculosis.

367 sheep and goat sera were analyzed by ELISA PPA3 (lyophilized protoplasmic protein strain Map 18) at a concentration of 0.04 mg/ml and cut-off point 0.80, these same sera were analyzed by MAA ELISA (proteins in suspension of a culture filtrate of Maa strain D4) at concentrations of 0.1 mg/ml, 0.2 mg/ml and 0.3 mg/ml. From the optical densities obtained in ELISA PPA-3 and Maa ELISA, by correlation and concordance tests, the concentration and cut-off point established for the use of Maa in indirect ELISA for the diagnosis of paratuberculosis were determined.

The correlation coefficient using the Sperman method was 0.84, where it was established that the highest agreement is 0.3 mg / ml and at a cutoff point 0.90 with 95% confidence intervals.

The use of the PPA-3 antigen is one of the most widely used worldwide and on which the diagnosis of paratuberculosis is based, but it is not widely available in Mexico; However, at the national level it is necessary to have an alternative for the use of another antigen, such as Maa, since its use generates lower cost and is available in the country without the need for its importation. The results obtained in the present study allow us to recommend the use of Maa ELISA as an alternative for the diagnosis of paratuberculosis and it is suggested as a screening test in places where the disease situation is unknown.
This study was supported by the project UNAM, DGAPA PAPIIT IT201118

1.22 DETECTION OF MYCOBACTERIUM AVIUM SUBSP PARATUBERCULOSIS THROUGH DIFFERENT DIAGNOSTIC TECHNIQUES IN UNGULATES KEPT IN CAPTIVITY IN MÉXICO.

Ana Laura Hernández Reyes[1], Gilberto Chávez Gris[1], Edith Maldonado Castro[1], Luz Elena Alcaraz Sosa[1]

[1]*Universidad Nacional Autónoma de México.*

Paratuberculosis is a chronic bacterial disease that affects domestics and wild ruminants, in the last decade studies in wild animals have increased, identifying the presence of M. paratuberculosis in wildlife, and their participation as possible sources of transmission to other species. In Mexico the status of paratuberculosis in wildlife is unknown, due to the above, it is important to carry out studies in order to know the health status of the disease.

The objective of this study is demonstrate the presence of *Mycobacterium avium* subsp. *paratuberculosis* (M. paratuberculosis) in an area with suspicion of this disease, where coexist 3 species of ungulates kept in captivity, giraffe (Giraffa camelopardalis) blue wildebeest (Connochaetes taurinus) and scimitar-horned oryx (Oryx dammah).

During eight months, fecal samples were obtained directly of the ground and intestine samples were collected from the ungulate mortality during the eight months. In the intestinal samples, mycobacterial concentration from the intestinal mucosa was collected using a modified version described by Ratnamohan and Spencer (1986), then the DNA from mycobacterial concentration was extracted using QIAamp DNA mini Kit (QIAGEN®) and from the fecal samples the extraction of DNA was according to the technique of Garrido et al., (2000). The IS900 PCR and a multiplex DMC PCR (Collins et al. 2002), were performed. The culture medium used for both samples was Herrold egg yolk medium with mycobactin and pyruvate. In the histopathological studies, the tissues preserved in formalin were processed and then stained with hematoxylin and eosin (HE) and Ziehl-Neelsen.

The total of fecal samples collected were 56, 20 of giraffe, 12 of blue wildebeest and 24 of scimitar-horned oryx, also a total of five samples of small intestine in the scimitar-horned oryx were collect. Only individual of scimitar-horned oryx were positive, a total of 6 samples, 4 faeces samples and 2 intestinal samples were culture positive, the time of colonies growth was at 8 to 10 weeks, the bacterial colonies of M. paratuberculosis were confirmed by IS900 PCR. A total of five samples, 2 faeces samples and 3 intestine samples, were IS900 PCR positive, in the DMC PCR an amplified corresponding to strain C (cattle), was obtained. The amplification obtained from the IS900 PCR of bacterial colonies was sequenced and compared in a data base of sequences (BLAST), obtaining a homology of 100% with other sequences of M. paratuberculosis. In the anatomopathological studies, 2 samples of intestine, showed thickening of the mucosa with congestion, histopathologically a diffuse infiltration of macrophages with abundant acid-fast bacilli in the lamina propria of small intestine were observed.

This study is the first of its kind in Mexico, where using more than one diagnostic technique and a non-invasive sampling method, the presence of M. paratuberculosis was demonstrated. This species is extinct in the wild since year 2000 (IUCN), its conservation depends on the population kept in captivity, for this reason maintain in a healthy state at this species allows to participate in conservation and reproduction programs. In adition this study could collaborate to realize research of paratuberculosis in free- ranging wildlife in Mexico. Research carried out thanks to the Program UNAM, DGAPA PAPIIT IT201118

1.23 THE POTENTIAL OF VOLATILE ORGANIC COMPOUNDS AS BIOMARKERS OF PARATUBERCULOSIS

Heike Koehler[1], Anne Küntzel[1], Sina Fischer[1], Elisa Kasbohm[2], Wolfram Miekisch[3], Petra Reinhold[1]

[1]*Institute of Molecular Pathogenesis, Friedrich-Loeffler-Institut, Jena*
[2]*Department of Mathematics and Computer Science, University of Greifswald, Greifswald*
[3]*Department of Anesthesia and Intensive Care, Rostock University Medical Center, Rostock*

The analysis of volatile organic compounds (VOCs) was considered an alternative diagnostic approach for paratuberculosis, first to speed up bacterial culture of MAP and second for direct on-farm detection of infected animals.

The aim was to identify methodological and biological factors that influence VOC profiles in the headspace above cultures and in headspace above feces and breath of animals to assess the potential of VOC analysis for the in vitro and in vivo diagnosis of paratuberculosis.

Bacterial cultures of up to five different MAP strains in serial twofold dilutions were incubated on HEYM and four other media all supplemented with mycobactin for 2, 4 and 6 weeks. VOCs in the headspace of these cultures were pre-concentrated by needle trap micro extraction or solid phase micro extraction and subsequently analyzed by gas chromatography-mass spectrometry. In two in vivo studies goat kids were orally inoculated with MAP suspended in milk replacer. Feces and breath of the MAP inoculated (n = 10-14) and control animals (10 each) were sampled in regular intervals over roughly one year. Headspace above fecal samples and alveolar breath were pre-concentrated and analyzed essentially as the headspace above bacterial cultures.

Cultures of MAP emit a characteristic panel of VOCs with minor differences between MAP-strains. Composition of the VOC profile varies depending on the culture medium used and culture time. VOC profiles in the headspace above fecal samples and in breath change with increasing age and with feeding (milk versus plant-based diet). The concentrations of particular VOCs are directly modified by feed intake. However, in the headspace above feces a panel of VOCs could be identified which was characteristic for MAP infection and allowed differentiation of infected and non-infected animals.

In conclusion, VOC profiles related to MAP and to paratuberculosis exist and can be detected in vitro and in vivo. As a next step, a core volatome has to be defined. For diagnostic application, further validation is necessary.

1.24 DETECTION OF MYCOBACTERIUM AVIUM SUBSP. PARATUBERCULOSIS IN FAECES AND TISSUE OF SMALL RUMINANTS USING A NON - AUTOMATED LIQUIDE CULTURE SYSTEM METHOD.

Luigi De Grossi[1], Davide Santori[1], Antonino Barone[1], Silvia Abbruzzese[1] Matteo Ricchi[2], Gaetana Anita Marcario[1]

[1]*Istituto Zooprofilattico Sperimentale del Lazio e della Toscana "M. Aleandri"*
[2]*Istituto Zooprofilattico Sperimentale della Lombardia e dell'Emilia Romagna "B. Ubertini"*

In Italy the strains of Mycobacterium avium subsp. Paratuberculosis in bovine are frequently isolated and studied while detection of MAP in sheep and goats is very rare and the circulating strains and their characteristics are unknown. The Paratuberculosis is very common in both species, it needs thereby to know more date about them for developing a control program.

The aim of this study was to detect strains of MAP type 1 (S) in ovine and goats using a cultural liquid manual method.

In this study, we investigated four flocks with Paratuberculosis and one flock considered to be free of Paratuberculosis in previous research. A total of 603 serum samples and 419 faeces samples were collected and serum samples were analysed by Enzyme- Linked Immunoadsorbent Assay (Paratuberculosis ELISA IDVET) Faeces samples were tested using Herrold's Egg Yolk Medium (HEYM), Middlebrook liquid medium (7H9+) and Real Time PCR (IS 900). During the experiment two positives animals died and tissues and faeces were investigate for MAP.

All four flocks were positive to MAP and one confirmed negative. 86 serum samples were positive to MAP by Elisa test.17 samples of faeces from the same subjects, almost with high value of % S/P (> 100), and 7 negatives were selected to be cultivated in liquid and solid medium and liquid culture analysed by RT PCR.Periodically, liquid cultural were tested by RT PCR to detect IS900 and positives were typed by F57.At last 28 samples, including faeces and tissues, were cultivated and 12 of these were positives at PCR IS 900 for MAP. 6 strains resulted type 1 (S) by PCR respectively from two flocks and four sheep.

Results obtained are important because represent one of the first ovine strains of MAP isolated in Italy and cultural method used was non automated and not expensive.

1.25 COMPARISON OF VERSATREK SYSTEM WITH SOLID HERROLD'S MEDIUM AND PCR METHODS FOR THE DETECTION OF MYCOBACTERIUM AVIUM SUBSP. PARATUBERCULOSIS (MAP) FROM BUFFALOES FAECES.

Pietrella Gabriele[1], Bartolomé del Pino Leticia Elisa[1], Cersini Antonella[1], Gamberale Fabrizio[1]

[1]Istituto Zooprofilattico Sperimentale del Lazio e della Toscana M. Aleandri

Due to the resistance of buffaloes to contract the disease, further studies using the Versatrek instrument to improve and validate an additional system to detect MAP in these species are under way.

This study aims at evaluating the reliability of the VersaTREK® system (Thermo Scientific) when compared to solid Herrold's medium and PCR methods used to isolate MAP.

Based on the results of a serological screening program obtained using ELISA (IDScreen® Paratuberculosis Indirect, Idvet), two farms, one negative (A) and one positive (B), were included in the study. From June 2015 to September 2016, 30 faecal samples from farm A and 66 from farm B were collected. The isolation on the Herrold's medium, DNA extraction (QIAamp® Blood Mini kit, QIAGEN), real time targeting IS900 (TaqMan®, Applied Biosystem) and isolation on VersaTREK® system were carried out on each sample. Positive cultures detected by the instrument were confirmed using real time F57 (TaqMan®, Applied Biosystem).

All samples from farm A resulted negative to HEYM culture and by IS900 real-time PCR. Only 1 positive culture by VersaTREK was detected, but it was not confirmed by F57 real-time PCR. 10 samples from farm B were flagged positive by VersaTREK: 4 out of them resulted positive to both methods (HEYM culture and real time IS900), 4 were confirmed positive by at least one test and 2 resulted negative when tested by other assays and their cultures resulted negative to F57 real time PCR.

These results suggest that VersaTREK system shows a fair sensitivity since several positive samples were confirmed by at least one method. This occurrence brings into question the specificity since the positive samples were detected only by the instrument and not confirmed by other tests.

1.26 ELISA UREA IN THE DIAGNOSIS OF PARATUBERCULOSIS: ESTIMATION OF SEROPREVALENCE IN SOME LOCALITIES OF BUENOS AIRES PROVINCE, ARGENTINA.

Maria Fiorella Alvarado Pinedo[1], Pedro S. Sosa[1], Leandro A. Di Paolo[1], Magali A. Romero[1], Luis M. Peralta[1], Gabriel E. Travería[1], R. Damian Moyano[2], Maria Jaureguiberry[3], M. Isabel Romano[3]

[1]*Centro de Diagnóstico e Investigaciones Veterinarias (CEDIVE), Facultad de Ciencias Veterinarias (FCV), Universidad*
[2]*Nacional de La Plata (UNLP), Buenos Aires, Argentina*
[3]*Instituto de Biotecnología, CICVyA INTA, Buenos Aires, Argentina.*
[4]*Cátedra y Servicio de Reproducción Animal, FCV, UNLP, Buenos Aires, Argentina.*

ELISA is the diagnostic test of choice in cattle with subclinical paratuberculosis. The incorporation of urea in ELISA has a chaotropic effect, assessing the strength of bound between antibodies and antigen, by mean of disrupting immune complex containing weakly bound antibodies with low avidity; the released antibodies are swept in the washing step, while the high affinity antibodies against Mycobacterium avium subsp paratuberculosis (Map) remains attached to microplate.

This work describes the results of 10.749 bovine serum samples belonging to 35 beef herds and 3 dairy herds carried out in 12 localities of Buenos Aires province.

After serum samples were incubated with antigen, plates were washed 3 times and 100?l of 8M urea solution was added to ELISA polycubates, and incubated 4 minutes at room temperature, following washing step, the rest of the procedure was continued as published elsewhere. The performance of this test was calculated based on the culture of 369 fecal samples, from animals with positives, suspicious and negative ELISA test results. For Map isolation, fecal samples were inoculated in liquid medium (M7H9C).

Individual seroprevalence for beef herds and dairy herds was 6% and 17%, respectively, however herds seroprevalence were 97, 8%. Positive isolations from the 369 fecal samples were successfully accomplished in 67% of positive ELISA urea animals, 54% of suspicious ELISA urea animals and 7% of negative ELISA urea animals. With Bayesian statistics the positive predictive value was 0, 61, and negative predictive value was 0,93.

These results are an update for regional seroprevalence demonstrating an increasing prevalence and confirm the utility of ELISA urea in the diagnosis and control of cattle with paratuberculosis.

1.27 COMPARISON OF DIAGNOSTIC METHODS IN PARATUBERCULOSIS INFECTED HERD

Iva Slana[1], Vladimir Babak[1], Radka Dziedzinska[1]

[1]*Veterinary Research Institute*

Causative agens of paratuberculosis (PTB), Mycobacterium avium subspecies paratuberculosis (MAP) can be diagnosed by several methods. These are direct (real time PCR, cultivation) and indirect (ELISA) detection methods. Its use depends on what we expect from the method. Whether we use the cheap, fast, sensitive method or only in this time the "fashion" method. On the base of the field data, the statistics can be used to determine the reliability and comparability of the individual diagnostic methods used in common veterinary practice.

The objective of this study was to compare results of three diagnostic method (ELISA, cultivation and real time PCR) applied onthe naturally infected cows in different stage of the MAP infection

Two medium and two small breeds were selected. In all four breeds PTB occurred, including occasional clinical forms. From all animals over the age of 18 months, faeces and blood were collected. DNA from faeces was isolated by the QIAamp DNA Stool kit according to the slightly modified manufacturer's instructions. For Cultivation the faeces were decontaminated and inoculated onto special solid growth media with supplement. Cultivation was done in 37 °C for three months. Blood samples were taken from serum. Antibody diagnosis was based on the ELISA method using the ID Screen Paratuberculosis Indirect Kit. The Cultivation, Real Time PCR, and ELISA results were statistically evaluated using Fisher's test and chi-square test. And statistical model was created.

Within the study, 650 animals originated from four cattle farms were tested for PTB. From all animals, 304 (46.8%), 109 (16.8%) and 76 (11.7%) were determined as positive using Real Time PCR, ELISA assay and culture, respectively. The results of all three methods start to correspond with each other in animals being in clinical phase of infection and shedding more than 104 of MAP in gram of faeces. Statistical analysis showed that the animal determined as positive in qPCR examination will be assessed as positive only in 35% using ELISA test. Using culture, the probability of positivity will be even lower (25%).

Statistical analysis showed that the animal determined as positive in qPCR examination will be assessed as positive only in 35% using ELISA test. Using culture, the probability of positivity will be even lower (25%). This work was supported by the MEYS CZ (NPU I LO1218).

1.28 COMPARISON BETWEEN TWO MOLECULAR GENOTYPING METHODS TO EVALUATE GENETIC DIVERSITY FOR MYCOBACTERIUM AVIUM SUBSP PARATUBERCULOSIS IN ARGENTINA

Brenda Vasini Rosell[1], Andrea Gioffré[1], Karina Cirone[1], Claudia Morsella[1], Laura Mendez[1], Monika Moravkova[2], Fernando Paolicchi[1]

[1]*INTA Castelar Biotechnology Lab*
[2]*Veterinary Research Institute Czech republic*

Paratuberculosis is a widespread intestinal disorder that causes chronic enteritis mostly in domestic ruminants. This disease is considered an important issue in Argentina, especially due to the high economic losses for farmers. It is caused by an intracellular pathogen called *Mycobacterium avium* subsp. *paratuberculosis* (MAP). Unfortunately there is no cure for this disease and an efficient vaccine is still on trials. Due to this and in order to improve control strategies of the disease we need a good understanding of the epidemiology; a useful tool for this purpose is strain differentiation using different multiple typing methods that have been developed.

The aim of this study was to compare two different methods for genotyping of MAP called Restriction Fragment Length Polymorphism (RFLP) and Multiple-Locus Variable number tandem repeat Analysis (MLVA) and evaluate the discriminatory power
(D) of each.

For the purpose of this study we genotyped a total of 26 MAP isolates belonging from cattle (n=21) and deer (n=5) and from ten (dairy and beef) herds from Argentina and provided by the strain bank of the microbiology laboratory in INTA Balcarce.

MLVA analyzed based on 8 loci revealed three different genotypes called INMV 1, INVM 13 and INMV 33 (names assigned by the National Institute of Agronomical Research in France (INRA)); only INMV 1 was represented in all the herds, and the other two appeared in only one herd. RFLP using BstEII restriction enzyme also showed three genotypes called C17, C1 and B. In this case the two less represented genotypes (C1 and B) appeared in two herds whereas C17 appeared in all the herds of the study. D was calculated with 12 non-epidemiologically related isolates for MLVA and 14 for RFLP and reached 0.3182 and 0.4835 respectively. Finally we evaluated D of the combined MLVA&RFLP methods taking into account all combinations found amongst MAP isolates studied. D reached 0.6083 when analyzing both methods together.

A comparison with other studies from European countries showed a low epidemiological association, as MLVA and RFLP most common patterns in Europe are INMV 2 and C1 respectively, while in this study no isolate with INMV 2 pattern was identified and C1 pattern appeared in 20% of the herds only. In regards of discriminatory power and according to the results of this study we conclude that the combination of both methods provides the greatest level of discrimination

1.29 MLVA GENOTYPING OF MYCOBACTERIUM AVIUM SUBSP. PARATUBERCULOSIS ISOLATES IN ARGENTINA

Brenda Vasini Rosell[1], Karina Cirone[1], Claudia Morsella[1], Laura Mendez[1], Andrea Gioffré[1], Fernando Paolicchi[1]

[1]*INTA Balcarce Microbiology Lab*

Paratuberculosis is a worldwide chronic enteric disease caused by Mycobacterium avium subsp paratuberculosis (MAP). In Argentina this disease cause considerable economic losses as it reduces the production of milk and meat, becoming an issue for the animal industry. Several methods to identify different strains of MAP have been developed; among them Mycobacterium Loci Variable Analyze (MLVA), consisting of eight loci has shown to be fast, easy and highly discriminatory; moreover, database from INRA in France presents a numerical profile called INMV.

The aim of the study was to describe the genetic diversity of MAP from different herds in Argentina by MLVA and evaluate the discriminatory power (D) of it.

A total of 91 isolates provided by the strain bank of the microbiology laboratory in INTA Balcarce were analyzed for the study. Reference strain ATCC 19698 was used as a control. Map isolates corresponded to cattle (n=86) and deer (n=5) and were obtained during the period from 1990 to 2016 from 25 herds all over the country.

The overall MLVA analysis yielded seven different MAP genotypes in Argentina; almost 75% of the isolates belong to a single dominant subtype called INMV 1 (n=68) which happens to appear in all the herds of the study. The rest of the genotypes were less represented, with INMV 2 (n=6), INMV 33 (n=5), INMV 3 (n=4), INMV 16 (n=2) and INMV 13 (n=1). The goeBurst analysis showed that INMV 2 is the founder genotype in the study. D was calculated with 37 non-epidemiologically related isolates and reached 0.536. With regards the D of each locus, loci X3, 3 and 32 showed no allelic diversity, whereas loci 292 showed the higher D value with 3 different alleles. We also studied the genetic diversity of the population and observed that six of the twenty-five herds analyzed presented more than one pattern (24%). All of these herds showed different patterns in the same year, confirming a coexistence of different genotypes in time and space. Moreover, a coinfection with two strains was observed in five isolates from two different herds. Genotyping of the strains from a mother and her fetus showed the same MLVA pattern (INMV 1) representing an evidence of vertical transmission of PTBC.

This study confirms that INMV 1 is the most frequent in the country and suggests a low epidemiological association with Europe where INMV 2 is the most popular. The coexistence of different strains in the herds strongly suggests the absence of animal monitoring prior to introduction. This represents a risk factor for infection of the herds.

1.30 DEVELOPMENT OF LOOP-MEDIATED ISOTHERMAL AMPLIFICATION (LAMP) ASSAY FOR DETECTION OF MYCOBACTERIUM AVIUM SUBSPP. PARATUBERCULOSIS INFECTION IN ANIMALS

Shivalingappa.Yamanappa Mukartal[1], Doddamane Rathnamma[1], Hogalagere Doddappaiah Narayanaswamy[2], Shrikrishna Isloor[1], Shoorvir Singh[3], Manju singh[3]

[1]Department of Veterinary Microbiology, Veterinary College, Hebbal, Karnataka Veterinary Animal and Fisheries Sciences University (KVAFSU), Be
[2]ICAR-NAE project, Department of Pathology, Veterinary College, Hebbal, Karnataka Veterinary Animal and Fi
[3]Division of Animal Health, Central Institute for Research on Goats (CIRG), Makhdoom, PO-Farah, Mathura-281122, Uttar Pradesh, India.

Johne's disease (JD) is chronic progressive enteritis in ruminants caused by *Mycobacterium avium* subsp. *paratuberculosis* (MAP). The disease is economically important in the livestock industry but its control is hampered by the lack of accurate rapid diagnostic tests. Range of diagnostic tests is available, but all have limitations.

In the present study, a loop-mediated isothermal amplification (LAMP) assay for the rapid detection of MAP was developed.

A total of 70 tissue samples were collected from pre-clinical and clinical cases of JD suspected cases of sheep from organized farm in Karnataka. The extracted DNA from samples were subjected to amplification by conventional PCR and LAMP targeting IS 900 gene. The LAMP assay was standardized using six primers to amplify IS 900 gene of MAP with various permutation and combinations in concentration of Magnesium sulphate, Betaine, dNTPs, temperature and incubation time. Amplification of MAP was optimized at concentration of 6 mM, Magnesium sulphate, 0.8 M Betaine, 1.4mMdNTPs at 63°C for 60min The successful amplification was indicated by a colour change from deep blue to light blue (Hydroxynapthol blue dye).

Among 70 samples, 88.50 (62/70) and 97.14 (68/70) per cent samples were positive by PCR and LAMP assay respectively. The sensitivity and specificity of LAMP assay for detection of MAP was 98.55 and 100 percent respectively.

To conclude the LAMP assay is simple to use, inexpensive, highly sensitive, and particularly well suited for the early diagnosis of Johne's disease in less well equipped laboratories

1.31 ISOLATION AND MOLECULAR CHARACTERIZATION OF MYCOBACTERIUM AVIUM SUBSPECIES PARATUBERCULOSIS FROM RUMINANTS IN KARNATAKA

Shivalingappa Yamanappa Mukartal[1], Doddamane Rathnamma[1], Hogalagere Doddappaiah Narayanaswamy[2], Shrikrishna Isloor[1], Shoorvir Singh[3], Kundan Kumar Chaubey[3]

[1]*Department of Veterinary Microbiology, Veterinary College, Hebbal, KVAFSU, Bengaluru,karanataka,India*
[2]*Department of Pathology, Veterinary College, Hebbal, KVAFSU, Bengaluru,karanataka,India*
[3]*Division of Animal Health, Central Institute for Research on Goats (CIRG), Makhdoom, PO-Farah, Mathura-281122, Uttar Pradesh, India.*

Johne's disease (JD) is a chronic incurable inflammation of intestines caused by *Mycobacterium avium* subspecies *paratuberculosis* (MAP) in ruminants.

In the present study domestic ruminants (sheep, cattle and goats) suspected for JD on the basis of clinical symptoms and pre-clinical cases were screened by multiple tests (direct microscopy, culture, ELISA, IS900, F57 PCR and IS1311 PCR-REA) for detection of MAP

A total of 237 faecal, 1474 sera, 236 blood, 192 tissue and six milk samples were collected from pre-clinical and clinical cases of JD suspected cases of cattle, sheep and goat from organised farm and farmers herd / flock from different geographical parts of Karnataka.

Among the samples, 294 (68.53%), 257 (17.40%), 541 (80.62%) 524 (78.09%) and 529 (78.83%) were positive by direct microscopy, ELISA, IS900 F57 and IS1311 PCR, respectively. A total 334 faecal and tissues samples screened by culturing in HEYM and Middle brook medium, 54 (32.33%) and 33 (19.76%) isolates were obtained respectively. All isolates were confirmed as MAP by microscopy and IS900 PCR. All the samples and isolates which were positive for IS1311 PCR were typed as 'Indian Bison Type' by IS1311 PCR-REA. The phylogenetic analysis of sequences confirmed that the isolates obtained by culture were 87 to 99 per cent homology with reference MAP-K10 strain. The per cent identity of the study isolates varied from 98.92 to 100 per cent.

The study indicated that MAP sequences of Southern India are related to MAP sequences of Northern India. Combinations of Microscopy, ELISA followed by PCR can be used for screening and diagnosis of JD in ruminants.

1.32 IDENTIFICATION OF BOVINE MICRORNA AS POTENTIAL BIOMARKERS OF EARLY MYCOBACTERIUM AVIUM SUBSP. PARATUBERCULOSIS INFECTION.

Gianpiero Zamperin[1], Elisabetta Stefani[1], Alice Fusaro[1], Massimo Bottazzari[1], Adelaide Milani[1], Annalucia Tondo[1], Alessia Schivo[1], Laura Paganini[1], Isabella Monne[1], Nicola Pozzato[1],

[1]*Istituto Zooprifilattico Sperimentale delle Venezie*

Effective control of the *Mycobacterium avium* subsp. *paratuberculosis* (MAP) is hampered by the lack of effective and accurate assays for its diagnosis. In particular, the time-gap between infection occurrence and clinical signs manifestation and the low sensitivity of current diagnostic assays make difficult to identify MAP during the sub-clinical phase of infection. Circulating microRNAs (miRNAs) have been shown to have significant potential as novel biomarkers for a range of human and animal diseases.

To improve early diagnosis of MAP infection through the identification of miRNA associated to the infected and infectious status of the disease

We followed 5 Holstein-Fresian herds in a 3-year prospective study where each animal was periodically tested for MAP infection by MAP-faecal culture, PCR and ELISA (734 samples from 478 animals). We investigated 40 samples from heifers and cows for miRNA identification by deep sequencing with the next-generation sequencer NextSeq500/550 and profiling using mirDeep2 software. For the differential expression analysis, 26 out of 40 samples were divided into five groups, selected based on animal age – young (10-15 months) or adult (>=23 months) – and disease status – infected, infectious and control.

Overall, we identified 408 known and 620 novel miRNAs among all samples analyzed. We found 6 known and 2 novel miRNAs as differentially expressed (DE). Specifically, all DE miRNAs were identified from the comparison adult-infectious vs adult control groups, 6 from adult-infected vs adult-control groups and 3 from young-infected vs young-control groups. All DE miRNAs showed decreased expression levels in control respect to infectious/infected animals and were involved in biological functions related to cancer, hematopoiesis, B-cells proliferation and generic immunology.

The eight known/novel miRNAs we identified in bovine blood may be used as potential diagnostic biomarkers of MAP infection and may complement the current MAP diagnostic tests for an earlier identification of MAP infected animals. Field evaluation by realtime- PCR is ongoing.

1.33 DEVELOPMENT OF RAPID ONSITE CASSETTE BASED SEROLOGICAL DIAGNOSTIC KIT FOR PARATUBERCULOSIS

Mukta Jain[1], Prudhvi Chand Mallepaddi[1], Rathnagiri Polavarapu[1], G. K. Aseri[2], Parul Yadav[2], Jagdip Singh Sohal[3]

[1]*Genomix Molecular Diagnostics Pvt. Ltd., Hyderabad, India*
[2]*Amity Institute of Microbial Technology, Amity University Rajasthan, Jaipur, India*
[3]*Amity Center for Mycobacterial Disease Research, Amity University Rajasthan, Jaipur, India*

Paratuberculosisis is a menacing disease of ruminants caused by *Mycobacterium avium* subsp. *paratuberculosis* (MAP). It leads to huge economic losses and has zoonotic concerns with Crohn's disease. Control of this disease is priority in developed world. However, in the developing world due to lack of diagnostic resources, control programs are yet to be initiated. In this respect development of affordable onsite rapid tests will promote the control programs in developing world. Onsite tests will not require laboratory infrastructure, expertise; and will avoid the transmission of the clinical materials thereby preventing the accidental exposure and cross contamination. Here we report the development of first onsite antibody detection kit based on lateral flow assay (LFA). We used recombinant culture filtrate protein antigens of MAP for developing the kit.

To develop lateral flow assay (LFA) based cassette for rapid onsite diagnosis of paratuberculosis

MAP culture filtrate protein antigens (CF041, CF040, CF341 and CF281) were expressed using pET151/D-Topo vector in E. coli. Kit was developed as compact plastic casing consists of a nitrocellulose membrane detection strip flanked at one end by a reagent pad and at the other end by an absorption pad. Detection strip contained 2.0 mg/ml of the each recombinant protein (at test line) as well as a reagent control line applied in distinct lines. Reagent pad contained colloidal gold conjugated Protein-G for antibody detection and streptavidin for control. For optimizing the test we used previously characterized positive (S/P ratio: ~1 in plate ELISA) and negative sera samples (149).

Results demonstrated the successful optimization of the onsite rapid LFA based antibody detection test. Test can diagnose the infection in just five minutes time. Of the tested 88 positive sera samples, 72 (81.8%) yielded positive reaction in LFA test, negative samples (61) remained negative.

Present study reports the first time development of onsite rapid LFA based diagnostic kit capable of diagnosing infection in 5 minutes, finding are enthusiastic and endorse the prospective field use of newly optimized onsite test. Farmer can directly use this kit without need of expert.

1.34 FIELD USE SKIN TEST BASED ON TYPE IV HYPERSENSITIVITY USING NOVEL SECRETORY ANTIGENS FOR DETECTING PARATUBERCULOSIS INFECTION

Mukta Jain[1], Rathnagiri Polavarapu[1], G. K. Aseri[2], Amit Kumar Singh[2], Deepansh Sharma[2], Jagdip Singh Sohal[3]

[1]*Genomix Molecular Diagnostics Pvt. Ltd., Hyderabad, India*
[2]*Amity Institute of Microbial Technology, Amity University Rajasthan, Jaipur, India*
[3]*Amity Center for Mycobacterial Disease Research, Amity University Rajasthan, Jaipur, India*

Paratuberculosis caused by *Mycobacterium avium* subsp. *paratuberculosis* (MAP) has been categorized as List B disease by OIE, a disease of both economic and public health importance. Control is largely dependent on diagnosis the infected individuals followed by culling or segregation. India is endemic for paratuberculosis, however, control programs are yet to be initiated. We lack indigenous diagnostics and imported kits are expensive. Herd screening tests, particularly onsite are preferred. Skin test based on intradermal inoculation of antigen followed by observation for late allergic reaction is one such onsite herd screening procedure. In spite of successful application of the skin test in the control of bovine tuberculosis, it is not popular for paratuberculosis because of poor specificity and sensitivity. We report here optimization of skin test using mixture of recombinant novel secreted antigens.

To develop skin test for onsite diagnosis of paratuberculosis using novel recombinant secretory antigens

Recombinant novel secretory proteins of paratuberculosis bacilli (CF041, CF040, CF341 and CF281) were expressed using pET151/D-Topo expression vector in E. coli. We selected apparently healthy (62, cattle- 32 and goat- 30) and suspected animals (48, cattle- 19 and goat- 29) based on physical and clinical (emaciation with persistent diarrhea) parameters. Animals were injected intradermally with 10 µg of each protein in PBS. Correct injection was confirmed by palpating a small pea-like swelling at each site of injection. European Union standards of skin thickness post inoculation (≥4.0 mm) were used to diagnose positive and negative reactors.

Positive reactions were observed using 10 µg of each antigen. Positive reactors developed skin thickness ranging from 4.2 to 13.8 mm. Of the 48 suspected animals, 36 (75%) were positive reactors, 22 (35.4%) of 62 animals of apparently health category also developed positive reaction.

Findings of the present investigation are promising and demands validation in the field. Positive reactions in apparently healthy animals may be due to wide presence of pathogen in environment and its regular exposure of host species to pathogen, this represents potential epidemiological risk. Once validated, this test has the potential to be used in control programs.

1.35 DEUTERIUM TRACER BASED CELLULAR DYNAMICS OF TISSUE MACROPHAGES ISOLATED FROM GOAT KIDS AND ADULT GOATS IN RELATION THE PATHOGENESIS OF CAPRINE PARATUBERCULOSIS.

Ad Koets[1], Antonios Zagaris[1], Lars Ravesloot[1], Karianne Lievaart-Peterson[2], Mariona Baliu-Pique[3], Sigrid Otto[3], Laura Ackermans[3], Kiki Tesselaar[3], José Borghans[3],

[1]*Wageningen Bioveterinary Research, Lelystad, The Netherlands*
[2]*GD Animal Health, Deventer, The Netherlands*
[3]*Laboratory of Translational Immunology, University Medical Center Utrecht, The Netherlands*

The cellular life span of tissue macrophages has not been previously considered as a factor in the pathogenesis of infections caused by intracellular bacterial pathogens such as *Mycobacterium avium* ssp *paratuberculosis* (MAP) which reside in macrophages. For slow growing bacterial pathogens such as MAP the life span of the cells they infect may be a critical factor determining the local bacterial load and infection progression.

We used oral supplementation of the stable non-radioactive hydrogen isotope deuterium in new born and adult goats to label DNA of dividing cells and estimate cellular life span.

At fixed time points during the deuterium upload and washout period animals were sacrificed and cells were isolated from various tissues (e.g. small intestine, lung and mesenteric lymph nodes). These cells were labeled using fluorescent monoclonal antibodies for phenotype and sorted by flow cytometry. In highly pure cell populations the amount of deuterated DNA was estimated using GC-MS. Data were fed into established mathematical models to estimate average cellular lifespans.

Results indicated that monocytes and macrophages isolated from tissues of young growing goats have an estimated average lifespan of 6-27days. In comparison macrophages isolated from adult goats have an estimated lifespan of 13-24 days. The different tissues of origin and to a lesser extend age appear to be explanatory factors concerning macrophage life span. Intestinal macrophages in young animals are predicted to have the shortest lifespan (on average 6 days).

With documented MAP bacterial doubling times of 24-48 hours in in vitro cellular systems the fact that intestinal macrophages are relatively short-lived cells indicate that cellular turnover may be a crucial factor next to dose of exposure in understanding the progression of infection. These data are currently modeled in within-host mathematical models of infection dynamics to add to our understanding of the pathogenesis of paratuberculosis.

1.36 SYSTEMIC MONOCYTE/MACROPHAGE CHANGES IN RESPONSE TO MAP INFECTION

Kumudika de Silva[1], Hannah B. Pooley[1], Shyamala Thirunavukkurasu[1], Hilary Connors[1], Chad Cooper[1], Douglas Begg[1], Auriol C. Purdie[1], Karren M. Plain[1]

[1]*University of Sydney*

Macrophages, the target cell for MAP infection, can be polarised into distinct functional phenotypes in response to signals from their microenvironment. M1 macrophages are microbicidal, pro-inflammatory and responsive to IFN-γ while M2 macrophages are poorly microbicidal and have anti-inflammatory properties. While it is known that MAP interferes with the protective mechanisms employed by macrophages, information regarding macrophage polarization in response to MAP exposure and its possible implications is scarce.

This study was undertaken to assess the functional state of macrophages in response to MAP infection both in vivo and in vitro.

Immortal cell lines as well as cultures of primary monocytes and tissues from cattle exposed to MAP were assessed using flow cytometry, gene expression and measurement of factors such as cytokines and nitric oxide.

In MAP-exposed and unexposed cattle, during the latent or subclinical stage of infection there was a heterogeneous macrophage activation pattern characterized by the presence of both classical (M1) and alternate (M2) phenotypes in blood. Exposure of macrophages to antigens from MAP led to variation in the production of nitric oxide, interleukin-10 and tumour necrosis factor α. In order to understand this phenomenon, in vitro studies were conducted to examine the effects of cytokines and MAP exposure on macrophage polarisation status. Additionally, macrophage localisation and activation in gut lesions from infected animals versus animals that have a protective immune response was examined to determine macrophage polarization states

The results indicate that there are changes in the activation state and responsiveness of monocytes/macrophages from MAP-exposed animals and aid in further understanding the local and systemic monocyte/macrophage changes in response to mycobacterial infections.

1.37 THE POTENTIAL ROLE OF TH17-LIKE IMMUNE RESPONSES IN JOHNE'S DISEASE POSITIVE COWS

Justin L. DeKuiper, Paul M. Coussens

Johne's disease (JD) is a chronic gastrointestinal disorder of ruminants caused by Mycobacterium avium subspecies paratuberculosis (MAP). To assist in MAP control efforts, we focus on understanding immune responses to MAP, defining correlates of protection, and improving diagnostic assays. Later stages of JD coincide with a classical Th2-like immune response. Defining importance of a classical Th1-like response in JD has been more difficult.

Indeed, mRNAs encoding the cytokines IL-23 and IL-17a are significantly elevated in PBMCs from MAP test positive (JD+) cows relative to PBMCs from test negative (JD-) cows after stimulation with MAP antigens. Both IL-23 and IL-17a production have been associated with Th17-like responses. Th17 cells are also defined by expression of IL-23 receptor (IL-23R).

To determine the relative prevalence of potential Th17 cells in PBMCs from JD+ and JD- cows PBMCs were isolated and analyzed by immunostaining and flow cytometry. Surface staining for T-cell type (CD4, CD8, TCR1 (Yδ T cell)) and IL-23R was performed after an 18-hour incubation with or without MAP

Fresh PBMCs from JD+ cows (n=12) contained a significantly higher proportion of IL23R positive cells than PBMCs from JD- (n=12) ($p<0.05$). However, ELISA results for IL-17a revealed higher concentrations of IL-17a secreted from PBMCs treated with MAP (n=20) than from PBMCs not treated with MAP (n=20), regardless of JD status ($p<0.0001$). Plasma from JD+ (n=20) cows revealed significantly less IL-17a circulating in the periphery than in JD- cows from two distinct sources (n=18 and n=20) ($p>0.05$ and <0.013, respectively).

This data suggests that Th17 cells may indeed play a role in immune responses to MAP infection and development or control of JD, however, IL-17 producing cells may localize to sites of infection. Additional work will focus on specific cell types producing IL-17a and IL-23 in response to MAP antigens and on the potential role of Th17-like cells in MAP infected tissues.

1.38 CATHELICIDINS REDUCE MYCOBACTERIUM AVIUM SUBSP. PARATUBERCULOSIS INTERNALIZATION IN MURINE MACROPHAGES

Karina Cirone[1], Priyoshi Lahiri[1], Ravi Holani[1], Yin Lin Tang[1], Jeroen De Buck[1], Eduardo R. Cobo[1]

[1]*Production Animal Health, Faculty of Veterinary Medicine, University of Calgary*

Mycobacterium avium subsp. *paratuberculosis* (MAP) paratuberculosis causes intestinal infections in domestic and wild ruminants. In cattle, MAP causes chronic enteritis, resulting in diarrhea and eventually death by emaciation. Critical in pathogenesis of paratuberculosis is invasion of MAP in macrophages and survival inside phagosomes by limiting fusion between phagosomes and lysosomes. Cathelicidins are recognized antimicrobial peptides and potential therapeutics.

The objective was to determine the potential of cathelicidins to reduce MAP invasion in macrophages.

Murine monocyte/macrophage cells (J774A.1) were infected with two MAP strains (wild type and K-10 expressing green fluorescent protein (GFP) from plasmid pWes4) for 3 and 24h. Intracellular identification of MAP in macrophages by confocal microscopy,gene expression of cytokines and host defense peptides in macrophages by real time qRT-PCR and protein determination of cytokines in macrophages by ELISA were assesed.

MAP quickly (3 h) invaded murine (J774A.1) macrophages, as confirmed by confocal microscopy using MAP expressing green fluorescence protein (GFP). Macrophages infected with MAP had increased transcriptional gene expression of pro-inflammatory Tnf-α, Il-1β and factor neutrophil attractant Il-8, but decreased Tlr4. Pre-treatment with synthetic human LL-37 cathelicidins prevented internalization of MAP into macrophages. The reduced MAP load was accompanied by diminished transcriptional expression of pro-inflammatory cytokines (Tnf-α, Il-1β, Ifn-γ) and a significant increase in Il-8, a chemoattracting factor for leukocytes. Treatments with cathelicidins reduced expression of genes for Toll Like Receptors (Tlrs) 2, essential for MAP invasion.

In an in vitro cell culture system, synthetic cathelicidins modulated Tlr-2, limited intracellular invasion of MAP into macrophages and suppressed production of tissue-damaging inflammatory cytokines.

1.39 PRIMARY ISOLATION RATES OF PARATUBERCULOSIS BACILLI ON DIFFERENT SOLID MEDIA

Amit Kumar Singh[1], Mukta Jain[1], Deepansh Sharma[1], Neeraj Khare[1], Parul Yadav[2], Jagdip Singh Sohal[3]

[1]*Amity Institute of Microbial Technology, Amity University Rajaasthan, Jaipur, India*
[2]*Amity University Rajaasthan, Jaipur, India*
[3]*Amity Center for Mycobacterial Disease Research, Amity University Rajasthan, Jaipur, India*

Isolation of *Mycobacterium avium* subspecies *paratuberculosis* (MAP) from infected individuals is the most vital for downstream studies. However, MAP is extremely fastidious organism and primary isolation is notoriously difficult. MGIT medium is recommended for primary isolation, but this medium is expensive and requires sophisticated instrument and is unaffordable for developing and under developed world. Present study aimed to investigate the comparative performance of three different egg based solid media (MB7H10, LJ and HEYM) in primary isolation of MAP from clinical samples of suspected animals.

To compare the efficacy of egg based solid media in primary isolation of paratuberculosis bacilli

Animals suspected for paratuberculosis were sampled either from farms in Jaipur District of Western India or slaughter house (Chainpura, Jaipur). Farm animals were sampled for fecal material and slaughtered animals for mesenteric lymph nodes (MLN) and intestine near ileo-cecal junction (ICJ). In total of 112 samples were collected (fecal- 62, MLN- 25 and ICJ- 25). Samples were decontaminated using 0.9% HPC and inoculated in triplicate on each of three egg based media (Herrold's Egg Yolk Medium- HEYM, Middlebrook 7H10 Medium- MB7H10 and Lowenstein-Jensen Medium- LJ). All media were supplemented with Mycobactin J and PANTA antibiotic mix, MB7H10 was also supplemented with ADC solution. Slants were observed for growth every 15 day interval up to one year. Colonies appearing were confirmed as MAP by ZN staining followed by IS900 PCR.

Out of 112 samples (fecal and tissues) tested, MAP was isolated from 59 samples (52.6%) (fecal- 37 and tissue- 22). Acid fast isolates having positive IS900 PCR reaction were only considered as MAP. Maximum isolations were recovered on MB7H10 medium, 47 (41.9%) samples, followed by LJ, 25 (22.3%) and HEYM, 07 (6.25%). Out of the positive isolations, 28 (47.4%) samples
were exclusively positive on MB7H10 followed by LJ, 08 (13.5%) and HEYM, 03 (5.08%). Out of 59 isolates, 39 (66.1%) grew only on a single medium (either MB7H10 or HEYM or LJ) and 20 (33.8%) sample grew on more than one medium. None of sample was positive on all three media and combination of MB7H10+LJ recovered, 16 (27.1%) followed by HEYM+MB7H10, 03 (5.08%) and HEYM+LJ, 01 (1.69%). For all type of clinical samples maximum isolations were observed on MB7H10 followed by LJ and HEYM. Growth pattern was also different for each medium. The colonies on HEYM are rougher, larger, and medium color also changed. Colonies on MB7H10 and LJ are small in size and smooth in appearance and uniformly distributed in whole culture tube.

Considering the exclusive isolation of MAP on all three media, it is recommended to use all three media in combination for primary isolation, however, if single media is to be used, MB7H10 should be medium of choice.

1.40 COMPARISON OF THE SENSITIVITIES AND SPECIFITIES OF FIVE ELISAS BASED ON DETECTION OF HOST BIOMARKERS FOR DETECTION OF MYCOBACTERIUM AVIUM SUBSP. PARATUBERCULOSIS (MAP) INFECTION

Cristina Blanco Vázquez[1], Marta Alonso-Hearn [1], Ana Balseiro[2], Ramón Antonio Juste[2], José Miguel Prieto[2], Rosana Torremocha[3], Beatriz soriano[5], Ricardo Ramos[4], Carlos Llorens[5], Javier Amado[6], Rosa Casais[2]

[1]*NEIKER, Instituto Vasco de Investigación y Desarrollo Agrario. Animal Health Department, Derio, Bizkaia, Spain*
[2]*SERIDA, Regional Service of Agri-Food Research and Development. Center of Animal Biotechnology, Animal Health*
[3]*Department, Deva, Asturias, Spain*
[4]*Parque Científico de Madrid, Unidad de Genómica, Campus de Cantoblanco, Madrid, Spain*
[5]*Biotechvana, Paterna, Valencia, Spain.*
[6]*LSAPA, Animal Heath Laboratory of the Principality of Asturias, Department of microbiology, Gijón, Asturias, Spain.*

The ELISA is the diagnostic method more frequently used to control paratuberculosis (PTB) in domestic ruminants. We have recently conducted transcriptomic analysis to identify prognostic biomarkers of PTB. However, no attempt has been made to apply biomarkers detection in serum samples as a diagnostic tool

The aim of this study was to compare the diagnosis value (semisum of sensitivity and specificity) of 5 ELISAs based on detection of selected bovine biomarkers (B1 to B5). Furthermore, the accuracy of these ELISAs was compared to that of a ELISA based on detection of MAP specific antibodies (IDEXX), MAP specific fecal bacteriological culture and PCR.

The sensitivity and specificity of each ELISA were estimated using sera from 27 culled Holstein cows from a Spanish farm. Animals were classified using as gold standard the presence (n=23) or absence (n=4) of histopathological lesions compatible with PTB in their gastrointestinal tissues. The cut off with the highest diagnostic value for each ELISA was selected by ROC analysis.

The ELISA based on detection of B1 and the IDEXX ELISA had the highest diagnostic values 0.69 and 0.71, respectively. The first had the highest sensitivity detecting animals with focal histopathological lesions (38.46%, 5 out of 13) while the latter had the highest sensitivity detecting animals with multifocal and diffuse histopathological lesions (80.00%, 8 out of 10). Both ELISAs combined had a global sensitivity of 56.52%, improving the global individual sensitivity of the IDEXX ELISA (43.47%) and the B1 ELISA (39.13%). In addition, the presence of B1 in the ileocecal valve of the 27 Holstein cows was also investigated by immunohistochemical analysis using a specific antibody. B1 was located in the Paneth and Globet cells of the intestinal microvilli

In conclusion, the use of the B1 and the IDEXX ELISAs combined might improve the diagnosis of MAP infections.

1.41 EFFECT OF PARATUBERCULOSIS INFECTION ON WELFARE PARAMETERS OF DAIRY COWS

Leo Simone[1], Calamari Luigi[1], Arrigoni Norma[2], Tamba Marco[2], Amadori Massimo[3]

[1]*Istituto di Zootecnica, Facoltà di Scienze Agrarie, Alimentari e Ambientali, Università Cattolica del Sacro Cuore, Piacenza, Italia*
[2]*Istituto Zooprofilattico Sperimentale della Lombardia e dell'Emilia Romagna, National Reference Centre for Paratuberculosis, Piacenza, Italia*
[3]*Istituto Zooprofilattico Sperimentale della Lombardia e dell'Emilia Romagna, Laboratory of Cellular Immunology, Brescia, Italia*

Paratuberculosis causes predisposition of cattle to metabolic disorders, mainly close to calving, but data on these aspects are often lacking.

The aim of this study was to evaluate the impact of paratuberculosis on metabolic conditions of infected cows compared with healthy cows reared in the same herd, around calving period.

A case-control study was performed: cases were positive to serum ELISA (IDEXX Paratuberculosis Ab test®) and/or fecal culture, controls were repeatedly test-negative and asymptomatic animals. Blood samples of 30 positive and 30 negative cows were collected 4 times, at days -30, +3, +10, +30, with respect to calving. The metabolic profile was measured in each sample, including positive and negative acute phase proteins (APP) (i.e. haptoglobin, bilirubin), as well as markers of oxidative status (i.e. Reactive Oxygen Metabolites -ROMt- and Ferric Reducing Ability of Plasma). Some APP values were used to calculate the Post-calving Inflammatory Response Index (PIRI), and the Liver Functionality Index (LFI) which define the inflammatory response during the first week and the first month of lactation, respectively. At each sampling point, BCS and hygienic score were also measured. Nine of the 30 positive cows were culled before 30 days post partum due the onset of symptoms.

No differences were found for LFI, BCS and hygienic score. Significant differences were found in Ca and albumin (lower in the positive group) and ROMt (higher in the positive group). Selecting cows positive to both tests (ELISA and fecal culture), ceruloplasmin became significant too (higher in positive group). PIRI, calculated at 3 days post-calving, was significantly different for the entire case-control group, and for the serological only positive subjects.

The lower PIRI values in the case group highlighted their altered metabolic profile. LFI did not differ significantly between groups, probably because the early culling of 9 cows decreased too much the cases number. A greater number of observations are needed though to corroborate the initial hypothesis. Funded by Italian Ministry of Health, PRC2013/016.

1.42 UBIQUITOUS ANTIBODY RESPONSES TO THE POLAR GLYCOLIPID PHOSPHADITYLINOSITOL MANNOSIDE (PIM) LIMIT SEROLOGICAL TEST SPECIFICITY IN CATTLE

Ad Koets

The accurate diagnosis of paratuberculosis in ruminant species remains challenging.The sensitivity of indirect diagnostic systems which measure a host immune response to infection, such as antibody based systems like ELISA and T cell response based systems such as interferon gamma release assays are affected by disease characteristics. Additionally the choice of antigen used to detect a host response to infection has a critical impact on test specificity. Many of the tests currently available rely on crude, partially or ill- defined antigen preparations from cultured mycobacteria such as tuberculins, protoplasmic antigens and ultrasonic extracts.

In the current study we focused on the presence of polar glycolipids (GL) in particular phosphaditylinositol mannoside (PIM) in crude mycobacterial antigen preparations and their potential role as antigens with diagnostic value.

Polar GL can readily be detected by ELISA and immune thin layer chromatography (TLC) in a number of antigenic preparations produced using different methods of extraction, protein precipitation and purification.

Antibody based tests using polar GL and sera from cattle with known infection status and non-infected controls provided evidence that specific IgG class antibodies are generated against these antigens in the majority of ruminant hosts but irrespective of infection status. Major immunological determinants on the polar GL are associated with mannose type carbohydrates on these GL which are common to many pathogenic and non-pathogenic (myco)bacteria.

We conclude that mycobacterial polar GL PIM present in commonly used antigenic preparations limit the diagnostic specificity of serological assays due to GL specific antibody responses commonly present in animals irrespective of their infection status with respect to paratuberculosis.

1.43 EVALUATION OF SEROLOGICAL TESTS FOR THE DETECTION OF PARATUBERCULOSIS IN ITALIAN BUFFALOS (BUBALUS BUBALIS): A CLASS LATENT APPROACH

Matteo Ricchi[1], Giorgio Galletti[1], Simone Russo[2], Fabrizio Gamberale[3], Esterina DeCarlo[4], Norma Arrigoni[2]

[1]*Istituto Zooprofilattico Sperimentale della Lombardia e dell'Emilia Romagna, Reparto di Sorveglianza Epidemiologica dell'Emilia-Romagna (SEER), Bologna Italy.*
[2]*Istituto Zooprofilattico Sperimentale della Lombardia e dell'Emilia Romagna, National Reference Centre for paratuberculosis, Podenzano (PC), Strada Faggiola 1, Italy.*
[3]*Istituto Zooprofilattico Sperimentale del Lazio e della Toscana 'M. Aleandri', Via Appia Nuova 1411, 00178 Roma, Italy.*
[4]*Istituto Zooprofilattico Sperimentale del Mezzogiorno, National Reference Centre for Hygiene and technologies of water buffaloes farming and production, Italy*

The ELISA test is the most used assay for paratuberculosis control and it is utilised in the Paratuberculosis Italian Guidelines for assigning the health ranking to cow and buffalo herds. Currently, available commercial ELISA kits for the diagnosis of paratuberculosis in buffaloes are not supported by robust validation data. The gold standard recommended by the OIE Terrestrial Manual for the in vivo diagnosis of paratuberculosis is the cultural assay. However, the sensitivity of these tests is low, not suitable to be used as standard in indirect test validation. Some researchers have proposed PCR tests as gold-standard because of its sensitivity and rapidity.

Aim of this work was the validation of commercial ELISA tests aimed at detecting paratuberculosis infected buffaloes. The project was designed to collect data from herds located in various areas with different prevalences.

So far, the sampling was carried out in Frosinone and Rome provinces from two different herds, one with a low and another with a higher prevalence. The blood and faeces of 449 buffaloes were analysed in parallel by Id-Vet ELISA test and IS900-qPCR. In order to evaluate the accuracy/performance of both tests, these preliminary data were analysed by a Bayesian two latent class model, combining different cut off for both tests. The model included strong prior only for performance of qPCR (Se 50%, 90% sure is between 40% and 60%; Sp 70%, 95% sure is higher than 50%).

Results showed that at each combination of cut off considered (ELISA 0.6 vs PCR 38, 36, 34 Cq; ELISA 0.7 vs PCR 38 Cq) Sp of both tests was very high, with Posterior Median always higher than 95% and narrow high density intervals (HDI). Conversely, the model failed in evaluating Se: for ELISA, the HDI are too wide, for PCR, the strong prior was not modified and, finally, the prevalences estimated by the model are similar, in contrast with what requested by the model.

We believe further samplings will improve the model.

1.44 IMMUNOLOGICAL EVALUATION OF RECOMBINANT BACTERIOPHAGE P35 AS A VACCINE IN A NATURALLY INFECTED SHEEP FLOCK

Victoria Elizabeth Castrellón Ahumada[1], Edith Maldonado Castro[1], María Elena Munguía Zamudio[2], Karen Manoutcharian[2], Antonio Verdugo Rodríguez[1], Yesmín María Domínguez Hernández[1], Gilberto Chávez-Gris[1]

[1]*CEIEPAA, Facultad de Medicina Veterinaria y Zootecnia de la Universidad Nacional Autónoma de México*
[2]*Instituto de Investigaciones Biomédicas, Universidad Nacional Autónoma de México*

Paratuberculosis is a chronic disease associated to *Mycobacterium avium* subsp. *paratuberculosis*. Vaccination strategies is a possibility to control this mycobacteriosis. P35 is a protein that could be used to immunization flocks affected with Jhone's Disease

This study was developed to evaluate the immune response in a flock affected with paratuberculosis using a bacteriophage P35

Two groups of sheep were established for the analysis of immunological behavior after application of the recombinant bacteriophage P35 as vaccine. Cytokines response analyzed after vaccination were GM-CSF, IFN-γ, IL-2, IL-4, IL-5, IL-10, IL-12 and TNF-α. Likewise, the presence of IgG in blood was evaluated as an indicator of the humoral immune response.

The immunological evaluation showed an increase in the types of response, Th1 and Th2, of the characteristics of TNF-α, IL-4, IL-10 and IL-12 which could be Indicator of types of immune response act together to limit the development of the characteristic clinical picture of paratuberculosis in animals infected with Map and vaccinated against PTB.

We conclude that vaccination against paratuberculosis using the recombinant bacteriophage generates an immunological response involving the profiles Th1 and Th2 expressed mainly from week 15 to week 54 in infected animals. We also suggest that this response may explain that sheep vaccinated with the recombinant bacteriophage P35 will not show lesions. Due to the abovementioned, it is proposed that this immunogen could be used vaccine component in the control of paratuberculosis in infected animals. This study was supported by the project UNAM, DGAPA, PAPIIT IT201118.

1.45 ISOTYPES CHARACTERIZATION OF THE MAP-SPECIFIC EARLY IMMUNE RESPONSE IN EXPERIMENTALLY INFECTED CALVES BY IMMUNOBLOT

Bárbara Fernández, Silvia Colavecchia[1], Ana Jolly[1], Fernando Paolicchi[2], Silvia Mundo[1].

[1]*Universidad de Buenos Aires, Facultad de Ciencias Veterinarias, Inmunología. Argentina*
[2]*Instituto Nacional de Tecnología agropecuaria, Departamento de Producción Animal, Unidad Integrada Balcarce. Argentina*

The understanding of the immune response in paratuberculosis (caused by *Mycobacterium avium* subsp. *paratuberculosis* (Map) is still limited by a lack of tools to diagnose Map soon after infection occurs.

The aim of this work was to characterize the early humoral immune response induced after experimental infection of calves.

Holstein calves (6-7 weeks old) were experimentally infected per os with 250 mg of total wet weight live Map belonging to two Argentinean isolated strains. The groups were: Map strain A infected (MA; n=3), Map strain C infected (MC; n=2) and mock infected (MI; n=3). Specific IgG, IgG1 and IgG2 in sera, against sonicated Map (ATCC 19698) were analyzed by immunoblot prior-infection (0) and 150 days post-infection (dpi).

IgG reactivity to antigens of lower molecular weight than 50kDa was detected in sera of all infected calves at 150 dpi. Two wide bands (~35-47kDa and ~21-30kDa) were detected in sera of MC group when total IgG was evaluated. IgG1 showed no or a weak reactivity in all calves. Meanwhile, immunereactivity by igG2 resulted in heterogeneous band patterns (<50kDa) in most of the infected calves (2/3 MA and 2/2 MC). But a band greater than 50 KDa could be detected only in infected animals of MA group.

Our results showed different reactivity patterns between strains, especially in the study IgG2 isotype. From our results, the study of antigens of lower molecular weight than 50kDa in combination with the detection of IgG2 isotype emerges as an interesting alternative to be further explored in order to improve early paratuberculosis diagnosis.

1.46 CYTOKINE GENE EXPRESSION PROFILES IN THE BLOOD OF CLINICAL AND NON-CLINICAL SHEEP AND GOATS INFECTED WITH MYCOBACTERIUM AVIUM SUBSPECIES PARATUBERCULOSIS

Ganesh Gangaram Sonawane[1], Bhupendra Nath Tripathi[2], L.L.L. Prince[3], Rajendra Kumar Meena[2].

[1]*ICAR-Central Sheep and Wool Research Institute, Avikanagar, Malpura, Distt-Tonk- 304501(Rajasthan), India.*
[2]*ICAR-National Research Centre on Equines, Hisar-125001(Haryana), India.*
[3]*ICAR - Directorate of Poultry Research, Rajendranagar, Hyderabad 500 030 (Telangana), India*

Paratuberculosis is a chronic, infectious disease of domestic and wild ruminants characterized by granulomatous enteritis, progressive emaciation and irreversible diarrhea. It is caused by *Mycobacterium avium* subspecies *paratuberculosis* (MAP) a hardy, slow growing acid-fast microorganism. The role of cytokines in pathogenesis of paratuberculosis, in sheep and goats has been rarely studied in India.

The present study was aimed to evaluate expression profiles of certain cytokine genes in the blood of MAP infected and non infected sheep and goats maintained in the intensive system of management.

A total of 385 adult sheep and 55 adult goats of the paratuberculosis endemic and non-endemic farms respectively, were screened for MAP infection using multiple diagnostic tests. On the basis of clinical symptoms, ELISA, PCR and faecal examination, these animals were divided in to different groups as MAP infected clinical form (ICF, 9.8% sheep), MAP infected non-clinical form (INCF, 5.9% sheep) and MAP negative (NI, 84.3% sheep). Amongst 55 adult goats 6 were (4.5%) in the clinical form (ICF) of the disease while remaining 49 goats (95.5%) were negative (NI) for MAP. The healthy sheep from non endemic farm, for paratuberculosis served as control animals (CS) for the study. These sheep were negative for MAP infection by multiple tests. Blood samples were collected from representative animals from each group and relative gene expression levels of TGF-β, IL-10, IFN-γ and TRAF-1 cytokines were evaluated by using qPCR.

In sheep, the comparison between various groups revealed significantly higher mean relative expression of TGF-β, IL-10 and TRAF-1 in the NI sheep as compared to the ICF, INCF and CS groups. The mean relative expression level of IFN-γ in NI and INCF sheep was found to be comparable; however, it was significantly higher than the ICF and CS sheep. In the ICF sheep, the expression levels of all the cytokines under study were found to be down regulated with significant decrease in IL-10 level. The comparison of various cytokines between NI and ICF goats did not exhibit significant difference in the mean relative expression levels of all the cytokines determined in the study.

From the present study, it can be concluded that the immunological response in the MAP negative sheep was stronger than MAP infected non clinical and clinical sheep of the paratuberculosis endemic farm. The trend of cytokine expressions due to MAP infection in sheep and goats in the present study indicate different immunological response in both the species and required further comprehensive studies including identification of resistance markers to map infection.

Session 2. Map Control Programs and Epidemiology

LECTURE: INTERNATIONAL CONTROL PROGRAMS FOR PARATUBERCULOSIS - WHAT, WHY, HOW AND WHERE?

RICHARD WHITTINGTON
Sydney School of Veterinary Science
The University of Sydney
425 Werombi Road, Camden NSW 2570 Australia

Control programs for paratuberculosis have existed since at least as early as 1920. The chronology was documented by Benedictus and others in 2000. Since then new control programs for paratuberculosis have been implemented, and two reviews were published covering the situation in some countries up to 2012, bringing information together mainly related to cattle. However, there are other livestock species and new programs have commenced while others have been discontinued or substantially changed and there is a lack of current, authoritative information. It is impossible to ascertain from any single source what is being done about paratuberculosis internationally and the reasons for action or inaction on this disease. Consequently, animal health authorities are not in a good position to make recommendations to their own governments or domestic animal industries.

In this study colleagues from more than 40 countries came together through a structured process to summarise programs for paratuberculosis for the period 2012-2018, to document the rationale for having these programs, to assess the outcomes of some of the past and current control programs in different countries, and to make recommendations for future control programs in the light of competing priorities in animal health. While analysis is ongoing, the results of the work to date will be presented for discussion at the 14ICP.

Oral

2.1 ASSIGNMENT OF MAP TYPE C STRAINS FROM GERMANY TO GLOBAL PHYLOGENETIC MAP GROUPS

Petra Moebius[1], Heike Koehler[1].

[1]*Institute of Molecular Pathogenesis, Friedrich-Loeffler-Institut, Jena.*

In former studies about 80 genotypes were found for MAP type C isolates from Germany by combining complementary genotyping techniques (IS900-RFLP, MIRU-VNTR and MLSSR). This approach has a very high discriminatory power but is quite an effort. Results are hardly comparable with data from other studies using individual methods exhibiting a lower rate of MAP discrimination.

The aim of this work was to investigate the distribution of German MAP type C genotypes within the global phylogenetic tree of MAP strains, which is based on SNP analysis of whole genome sequences of isolates from different countries worldwide.

84 MAP type C isolates exhibiting 38 different combined genotypes originating from 9 host species and 9 Federal States of Germany were characterized by a novel SNP-based assay (Leao et al., 2016). Using this method, MAP type C isolates can be assigned to 14 phylogenetic groups by PCR and restriction enzyme digestion or sequencing of amplification products.

About 75% of the German isolates belonged to the main Sub-group A and 25% to Sub-group B. Most isolates of Sub-group A are distributed to 4 clades with SNP profiles 4, 6, 8, and 10. Strains within these clades were sub-differentiated by multi-target genotyping. Subgroups and clades consist of specific combined genotypes without overlapping. Independent from geographical origin, identical combined genotypes belonged to identical phylogenetic groups. Epidemiological links revealed by combined genotyping in a restricted area around National park Eifel were also confirmed by the new SNP-based assay. However, seven strains could not be classified by the current SNP panel. These strains and isolates of Sub-group B will be characterized by SNP analysis of additional WGS data to augment the panel of SNPs for unambiguous MAP classification.

The study shows the phylogenetically high diversity of German MAP type C strains in a global scale. Results will contribute to discover new SNPs for providing a phylogenetically robust framework for strain differentiation.

2.2 BREEDING FOR RESISTANCE AGAINST PARATUBERCULOSIS: HIGH GENETIC CORRELATION BETWEEN ANTIBODY RESPONSE AND FAECAL SHEDDING.

Lydia C.M. de Haer[2], Gerben de Jong[1], Maarten F. Weber[2].

[1]*CRV Arnhem, the Netherlands*
[2]*GD Animal Health, Deventer, the Netherlands.*
(Presenting author; *m.weber@gdanimalhealth.com).*

The heritability of antibody responses against *Mycobacterium avium* subsp. *paratuberculosis* (Map) has been estimated at 0.03 to 0.20, suggesting that breeding for resistance by selection on antibody response would be feasible. However, selection of breeding stock based on antibody response is attractive only if it results in offspring with a higher resistance to infection resulting in reduced shedding of Map, rather than in offspring being unable to mount an antibody response given infection.

Therefore, the aim of this study was to estimate the genetic relation between antibody response and faecal shedding.

Two data sets consisted of results of laboratory tests performed by GD Animal Health on samples from Dutch dairy herds. The first data set (PA1) consisted of 517,672 individual milk samples of 109,213 cows from 5,938 herds tested by ELISA for antibodies against Map between 2007-2010. The second data set (PA2) consisted of test results of 78,604 individual faecal samples of 52,348 cows from 435 herds. Faecal samples were tested between 1996-2015 by either modified Lowenstein-Jensen culture method, ESP- TREK culture system or qPCR assay. Heritabilities and genetic solutions for sires were estimated with a sire-maternal grandsire model with random permanent environment effect and with fixed effects herd*year, parity, birthyear and lactation period. In addition, for PA1 a covariable for milk production was included in the model, for PA2 a fixed effect for test method. Sire solutions were used to estimate MACE correlations (with correction for reliability) between PA1 and PA2. Sires with at least 15 daughters on 10 herds per trait were included in the evaluation, resulting in 446 sires for PA1 and 272 sires for PA2.

Heritability (se) for PA1 was 0.05 (0.003) and for PA2 0.06 (0.008). Repeatabilities for PA1 and PA2 were 0.42 (0.003) and 0.28 (0.006), respectively. The genetic correlation between PA1 and PA2 was 0.81.

This study confirms previous heritability estimates. Moreover, the high genetic correlation between traits for antibody response and faecal shedding indicates that selection based on antibody response is likely to result in offspring with a higher resistance to infection resulting in lower faecal shedding. In conclusion, this study indicates that breeding for resistance to paratuberculosis by selection on antibody response is feasible

2.3 ECONOMIC IMPACT OF CONTROL OPTIONS FOR JOHNE'S DISEASE IN CANADA.

Philip Rasmussen[1], Zhaoxue Ci[1], David Hall[1].

[1]*University of Calgary.*

Johne's disease (JD), also known as bovine paratuberculosis, is thought to be widely present within dairy herds in many countries including Canada. However, Canada is one of the few to have no mandatory control program in place. Starting with a review of control options for JD, we estimated the economic impact of using vaccination to control the disease in Canada.

Firstly, we aimed to summarize and evaluate existing control options for JD, and secondly, we aimed to simulate the adoption of a herd-level vaccine in Canada to estimate the economic impact of this control method.

We distributed a questionnaire to dairy producers across the largest dairy producing regions in Canada, obtaining farm-level demographic and production data. We then used a Monte Carlo simulation to estimate the economic impact of a vaccine being adopted, framing the resulting increase in milk production as a decrease in the cost of producing milk (i.e., improving production efficiency).

Many of the models reviewed relied on aggregated regional data, leading to significant variation in estimates of JD's economic impact and the benefits associated with different control methods. We addressed this by obtaining farm-level data for our simulation. Using assumptions of a 15% within-herd prevalence, an 85% level of willingness-to-adopt vaccination, and 10% greater milk production for healthy cows over diseased ones, we estimated the yearly economic benefit of controlling JD in Canada through vaccination to be approximately CAD$100 per cow.

Using more robust data than has previously been available to address Canadian willingness-to-adopt a bovine JD vaccine, as well as to model dairy production in Canada, we demonstrated a positive net benefit from adopting herd-level vaccination to control JD. Our research will contribute to the ongoing development of an effective control policy for JD in Canada and other countries

2.4 AN EFFECTIVE CONTROL PROGRAM USING A POOLED FAECAL REAL-TIME PCR ASSAY IN HERDS WITH JOHNE'S DISEASE.

Satoko Kawaji[1], Reiko Nagata[1], Akiko Mita[2], Makoto Osaki[1], Yasuyuki Mori[1].

[1]*National Institute of Animal Health, NARO*
[2]*National Livestock Breeding Center*

In a herd infected with Johne's disease (JD), it is assumed that the number of infected animals at the subclinical stage are several times more than that in the clinical stage. Control strategies in infected herds often involve test-based culling of infected animals supported by prevention of transmission. One of the key elements in this strategy is to detect infectious animals.

As an alternative to serological tests which are mainly used for the screening of whole herds even though lacking sensitivity, a novel ResoLight-based real-time PCR assay (RL-PCR) with pooled faecal samples was previously reported. In this study, JD infected herds were longitudinally screened with RL-PCR and antibody-ELISA, and the effectiveness of the control program was evaluated.

Eight JD infected herds (herd size 80-650) were selected. Screening of whole herds was conducted using RL-PCR and ELISA at intervals of 3-12 months for 2 years. For the RL-PCR assay, up to 10 individual faecal samples were pooled and tested. If the result of pooled RL-PCR was positive, faecal samples in the pool were tested individually by RL-PCR to identify positive animal(s). Animals that had positive results in either screening test were confirmed by a direct faecal real-time quantitative PCR (QPCR) assay, and all the animals diagnosed with JD were culled.

A total of 7100 animals were tested. Fourteen animals from 5 herds were diagnosed with JD through the screening test using RL- PCR followed by the QPCR assay, while only one animal was detected by ELISA. The RL-PCR assay with pooled faecal samples reduced the total number tested to 22.9% of individual testing, although the reduction per herd depended on the number of positive pools in the herd. At the initial screening of each herd, the percentage of RL-PCR positive pools ranged from nil to 75%, and it decreased over time in most of the herds.

The control program based on herd screening by RL-PCR and culling of reactors was effective in reducing the number of infectious animals.

2.5 A WEB BASED PARATUBERCULOSIS RISK ASSESSMENT AND MANAGEMENT SYSTEM FOR DAIRY FARMERS.

Dick Sibley BVsc HonFRCVS[1], Pete Orpin BVSc MRCVS[1].

[1]*Park Vet Group*

A paratuberculosis management program has been successfully used on many UK dairy herds using a predict and prevent model where risks of MAP entry and risks of disease spread within the herd are assessed to identify potential transmission routes which are then managed to minimise future prevalence

The study looks at practical ways to implement effective paratuberculosis control on infected farms

A web based health management tool has been developed to manage paratuberculosis on farms using a risk assessment and management plan (RAMP) generated by the program. The risk assessment tool identifies both biosecurity and biocontainment risks, managing the multiplier of disease. The system enables users to manage the identified risks practically and effectively to protect the

The program has been used on 1435 infected dairy herds to identify risks and define a control strategy.1293 opted for some form of risk assessment and management plan (RAMP) to control paratuberculosis. The program analyses risk and predicts future prevalence using an algorithm based on the current prevalence and the risks of spread.86% of herds had predicted increases in future prevalence if management practices at the time of the assessment were continued. Effective management plans were generated by the program.54% of herds use strategic testing to identify high risk cattle and focus required procedures on the high risk cows. Cow to calf transmission risks were identified as the main risk factor needing management

The program has delivered practical, effective and economic management plans for the control of paratuberculosis on many UK dairy farms with the principles forming the basis of a National Johnes Disease managementinitiative

2.6 FIELD VALIDATION OF DIVA ASSAY FOR INACTIVATED PARATUBERCULOSIS VACCINE.

Sujata Jayaraman[1], Mukta Jain[1], Kundan Kumar Chaubey[2], S. V. Singh[2], G. K. Aseri[1], Jagdip Singh Sohal[3].

[1]*Amity Institute of Microbial Technology, Amity University Rajasthan, Jaipur, India.*
[2]*Microbiology Laboratory, Central Institute for Research on Goats, Mathura, India*
[3]*Amity Center for Mycobacterial Disease Research, Amity University Rajasthan, Jaipur, India*

Paratuberculosis is a major production disease of ruminants with zoonotic concerns. It is chronic granulomatous enteritis leading to protein loosing enteropathy. Control of paratuberculosis in animals has become a priority due to farm economics and public health concerns. Recent research has proved that vaccination is the most practical method to control this disease. However, in the absence of DIVA (Differentiation of Infected & Vaccinated Animals) tool vaccination cannot be used because it interferes with diagnosis of paratuberculosis as well as tuberculosis. There are regulatory restrictions on mass use of paratuberculosis vaccines. Therefore DIVA test is needed to be developed. Commercially available vaccine in India is killed preparation of MAP cells. Therefore we developed DIVA using secretory antigens that are not part of vaccine but are potent antigens during natural infection.

To do the field validation of DIVA tool for inactivated paratuberculosis vaccine

Field validation of the previously optimized DIVA ELISA (Sohal et al., 2016) in our lab was done. In laboratory settings this prototype DIVA ELISA successfully differentiated vaccinated and naturally infected animals based on S/P ratio (Sohal et al., 2016). Vaccinated animals have S/P ratio ≤0.4 and infected animals have S/P ratio >0.4 in lab scale optimization. This DIVA ELISA utilizes four proteins in the secretome of paratuberculosis bacilli (MAP1693c, MAP3547c, MAP4308c and MAP2677c). Proteins were expressed in E. coli system using pET151/D–TOPO vector. For the field validation new vaccination trials were setup in Northern (District- Mathura) and Western (District- Jaipur) India using the whole cell inactivated vaccine in ruminant species at four different farms. These four farms were different management practices (intensive or semi-intensive or extensive). Total 39 animals (cattle- 14, goat- 24 and buffalo- 01) were included in the study. These animals were negative for paratuberculosis in ELISA and fecal shedding. Animals were monitored for a period of one year using DIVA ELISA, MAP PPD based ELISA and Fecal PCR at 0 DPV, 60 DPV, 120 DPV, 180 DPV and 360 DPV.

Prototype ELISA successfully differentiated infected and vaccinated animals in the field. Of the 39 vaccinated animals, 37 had S/P ratio ≤0.4 during the monitoring period. Two animals (cattle) had S/P ratio >0.4 after 60 DPV till the monitoring period, these animals were found positive for fecal shedding of MAP in PCR. These two animals belonged to the farm with poor hygienic management. None of the other animals (37) were positive for fecal shedding in PCR. All vaccinated animals had S/P ratio >0.4 in MAP PPD based ELISA.

Findings of the present study confirm the ability of DIVA ELISA in differentiating animals with respect to vaccination and natural infection. Study also shows that vaccination does not completely eliminate the chances of infection, therefore, hygienic management with vaccination is recommended for better results

2.7 PROTECTIVE LIVE ATTENUATED AND NANO-VACCINES AGAINST JOHNE'S DISEASE.

Adel M. Talaat[1], Akanksha Thukral[1], Chungyi Hansen[1], Kathleen Ross[2], Balaji Narasimhan[2].

[1]*University of Wisconsin-Madison.*
[2]*Iowa State University*

Mycobacterium avium subsp. *paratuberculosis* (M. paratuberculosis) causes Johne's disease in ruminants and is characterized by chronic gastroenteritis leading to heavy economic losses to the dairy industry worldwide. The currently available vaccine (inactivated bacterin in oil base) is not effective in preventing pathogen shedding and is rarely used to control Johne's disease in dairy herds.

To develop a better vaccine that can prevent the spread of Johne's disease, we utilized both live attenuated (LAV) and polyanhydride nanoparticles (PAN) of whole cell lysate (PAN-Lysate) or culture filtrate (PAN-Cf) of M. paratuberculosis.

Different mouse groups were immunized with both LAV and PAN vaccine constructs. All vaccine constructs were well tolerated in mice causing no inflammatory lesions at the site of injection.

Immunological assays demonstrated a substantial increase in the levels of antigen-specific T cell responses in the pre-challenge PAN-Cf vaccinated group as indicated by high percentages of triple cytokine (IFN-γ, IL-2, TNF-α) producing CD8+ T cells. Following challenge, some vaccine constructs (e.g. PAN-Cf) continued to produce significant levels of double (IFN-γ, TNF-α) and single cytokine (IFN-γ) secreting CD8+ T cells compared to inactivated vaccine group. A significant reduction of bacterial load was observed in the all mice organs of LAV and PAN-vaccinated mice, a clear indication of mice protection.

Overall, both LAV and PAN vaccines provided an attractive approach for developing protective and prolonged immunity against Johne's disease, an approach that could be applied for other intracellular pathogens.

2.8 *MYCOBACTERIUM AVIUM* SSP. *PARATUBERCULOSIS* (MAP) MOLECULAR DIVERSITY IN LATIN AMERICA AND THE CARIBBEAN: A SYSTEMATIC REVIEW.

Nathalia M Correa-Valencia[1], Miguel Hernández-Agudelo[1], Jorge A Fernández-Silva[1].

[1]*Centauro, Escuela de Medicina Veterinaria, Facultad de Ciencias Agrarias, Universidad de Antioquia, Colombia*

Little is known about MAP molecular diversity in Latin America and The Caribbean.

Therefore, we aimed to systematically collect and appraise the scientific evidence to answer the question: What MAP genotypes have been isolated from cattle, sheep, and goats in Latin America and the Caribbean?

An electronic search was conducted on two search platforms (OVID® and Web of Science®) including ten databases. The proceedings of the ICP held between 1991 and 2016 were also reviewed searching for evidence of further publications. Only articles published in peer-reviewed journals in any language were considered. The reference lists of relevant papers were searched for additional material. Screening by title, abstract, and full-text by two authors was performed.

A total of 522 publications matched the search terms and 24 articles met the inclusion criteria. All were published in English, in 13 different journals, and between 1989 and 2017. The relevant publications reported the use of six different genotyping techniques (PCR-REA, RFLP, IS900 type-specific-PCR, MIRU-VNTR, MLSSR, and SNP). Isolates were from Argentina (n=18), followed by Chile (n=11), Venezuela (n=4), Mexico (n=5), Brazil (n=2), and Colombia (n=3). The isolates were from cattle (n=14), goats (n=10), and sheep (n=4), and from feces (n=5), tissue (n=4), milk (n=3), drinking water (n=1), and pit slurry (n=1). One study reported the use of IS900 PCR-REA, six of IS1311 PCR-REA, one of IS1245 RFLP, and 10 of IS900 RFLP, using the enzymes BstEII, PstI, HinfI, MseI, PvuII, and BclI. MIRU-VNTR was used in eight studies, considering 17 loci, where the most frequently used were 1658 (X3), 32, 292, 10, 25, 47, and 3. MLSSR was reported in four studies, considering loci 1, 2, 8, and 9. SNP and IS900 type-specific-PCR techniques were reported by one study each. Genotypes found so far in the region using typing techniques were mainly C type. MIRU-VNTR mostly reported INMV 1, INMV 2, and INMV 11 subtypes. MLSSR reported by one publication, including genotypes from four different countries, found seven different subtypes of which 7g–10g–4ggt was the most common for loci 1, 2, and 8, respectively.

The high diversity of techniques used so far to genotype Latin-American and Caribbean MAP isolates makes difficult to answer the original question of this systematic review. However, we identified a relative genetic similarity between MAP strains recovered from cattle, goats, and sheep, regardless of the matrix and geographic origin

2.9 PHYLOGENETIC ANALYSIS AND EPIDEMIOLOGIC MODELING OF THE EFFECT OF *MYCOBACTERIUM AVIUM* SUBSP. *PARATUBERCULOSIS* GENOTYPE ON MILK PRODUCTION ON THREE MINNESOTA DAIRY FARMS.

Jonah Cullen[1], Yuanyuan Wang[1], Juan Abrahante[1], Fernanda Shoyama[2], Srinand Sreevatsan[2], Scott Wells[1]
[1]University of Minnesota.
[2]Michigan State University

Optimal herd-specific methods to control Johne's disease within dairy herds are not currently available, in part due to lack of understanding of multiple potential routes of transmission of the causative agent Mycobacterium avium subsp. paratuberculosis (MAP). Use of newly developed whole genome sequencing methods offers the potential to improve disease control through distinguishing infected cattle by genotype, to allow improved understanding of transmission routes. Moreover, differences in clinical disease presentation and lactation performance associated with MAP genotype may exist within herds.

An objective of this pilot study was to combine epidemiologic and whole genome sequencing data to assess associations between MAP genotype and production at three non-epidemiologically linked Minnesota dairy farms in 2014 and 2015.

Cows within three farms were tested for MAP (via serum ELISA and real-time fecal PCR assay) at two time points. A subset of cows with both positive serum ELISA and real-time fecal PCR results were selected for MAP whole genome sequencing. MAP was enriched directly from fecal samples using a peptide magnetic bead protocol for DNA extraction and direct de novo genome sequencing (Illumina MiSeq platform). Reads were mapped to the MAP K10 reference with the Burrows-Wheeler Aligner and SNP variants were called using FreeBayes. Hierarchical clustering was conducted using Unweighted Pair Group Method with Arithmetic Mean and a Jukes Cantor substitution model.

A total of 515 SNPs across 45 MAP samples were included in the analysis. For this study, genotype was defined by stratifying the phylogenetic tree at approximately the same branch length into four clusters. At least three of the four genotype groups were identified on each of the three study farms. While results did not detect genotype differences by quantitative serum ELISA result or fecal PCR cycle time, milk production differences (based on 305 day mature equivalent milk) were detected by genotype adjusted by farm.

These results suggest genomic differences (as assessed via phylogenetic relatedness) between MAP isolates may impact production- level outcomes. Genotype-specific control of MAP within farms could ultimately prove of economic value, indicating the need for future research examining strain diversity and within-herd transmission dynamics.

2.10 LONG TERM RESULTS OF AN EXPERIMENTAL VACCINATION TRIAL IN DAIRY CATTLE.

Joseba M. Garrido[1], Natalia Elguezabal, Ramón A. Juste[2], Miriam Serrano, María V. Geijo, Elena Molina, Iker A. Sevilla.

[1]*Neiker-Instituto Vasco de Investigación y Desarrollo Agrario. Departamento de Sanidad Animal, Derio, Bizkaia, Spain.*
[2]*Serida, Servicio Regional de Investigación y Desarrollo Agroalimentario de Asturias, Villaviciosa, Asturias, Spain.*

Johne's disease is not included in national control programs for cattle in Spain but it compromises the viability of herds. In order to evaluate an inactivated vaccine an experimental field trial has been in follow-up for more than 13 years in the Basque Country, Spain.

With the aim to control the disease, a program comparing two strategies, testing & culling (TC) and vaccination (VH), was setup in some volunteer herds from the Basque Country.

Starting in 2005 with three vaccinated (VH) and two test & cull (TC) Friesian herds, currently the trial includes 9 TC and 19 VH herds. In the VH herds, all animals older than 2 months are vaccinated with an inactivated vaccine at the time they join the program and from then on, only the replacers between 2 and 6 months old. Fecal and blood samples for PCR and ELISA, respectively, from animals older than 24 months are collected annually.

The initial average shedding prevalence in VH and TC herds was 12.63% and 8.27%, respectively. A sharp decline in the proportion of shedders was observed in both groups between the first and the second yearly sampling (YS). This trend continued in VH until the 6th YS when the proportion of shedders was 0.85%. In contrast, this figure settled at around 4% in TC herds during the same period. After vaccination, only 1.98% of the total number of animals included in VH herds were shedding MAP during their lifetime. However, for animals vaccinated when younger than six months of age the figure further decreased to only 0.48 (two months old) and 0.98% (between 2 and 6 months old). At the moment 12 herds have been subjected to more than 6 YS in the VH group and 8 (66.7%) out of 12 of these herds have no shedders since their 5th YS. Four (50%) of them have not had any positive animal in the last five samplings. In the TC group, in contrast, only one herd had negative results for 3 consecutive samplings but this was followed by one positive animal one year afterwards.

In conclusion, even though both strategies have reduced paratuberculosis prevalence, the economic costs associated to TC herds were much higher because testing and slaughtering ELISA or PCR positive animals. Therefore, this field trial shows that vaccination is a highly efficient strategy for paratuberculosis control both in epidemiologic and economic terms. It demonstrates that paratuberculosis eradication can be achieved at a reasonable cost, without extreme management changes, in a short period of time and consistently maintained afterwards.

2.11 ARE CATTLE INFECTED WITH MULTIPLE STRAINS OF MAP? A NEW COMPUTIONAL METHOD TO DETECT FROM WHOLE GENOME SEQUENCING DATA.

Yuanyuan Wang[1], Scott J. Wells[1].

[1]*University of Minnesota, Twin Cities*

At least 90% of U.S. dairy operations have Johne's disease (JD) infected cattle. High herd prevalence, extensive incubation period, and persistence of Map in farm environments have resulted in a complicated situation where a cow can be infected by multiple cows carrying distinct strains. However, no computational methods are available to identify multi-strain infected cases of Map from WGS data. Ignoring these special cases when constructing phylogenetic trees can rule out meaningful contact links between infected hosts. And, as strains compete for survival within a host, multi-strain infected animals are associated with increased transmission intensity, so identifying them is crucial for disease control.

Using whole genome sequences recovered from feces of infected cattle, we developed a computational method to identify multi- strains infected animals and quantify strain proportions.

A conceptual analogy is to determine multiracial individuals in a mixed population from their DNA profiles in two steps: 1. assemble a reference panel representing distinctive features of each race; 2. compare each profile with the reference panel to compute match percentage, and normalize the percentages as proportions. Similarly, we first selected several ancestor strains using allele frequencies derived from data. Then we quantified strain proportions using 2 techniques in data mining: non-negative least squares (aka dynamic programming) and non-negative matrix factorization.

Tested on 112 samples from 5 US dairy herds in Minnesota, our method identified ~10% animals infected with more than one strains. Recognizing these cases have resulted in structural changes of phylogenetic tree which altered transmission pathway in the epidemiological inference (who-infected-whom) of these farms. To evaluate the performance of our method, we used both synthetic sequencing data and in vitro lab mixtures of Map with known strain proportions of DNA. Knowing the ground truth of these strains, we are evaluating sensitivity (correctly identify mixed infections) and specificity (correctly identify single infections).

This work represents the first attempt to use WGS data to computationally quantify multi-strain MAP infections to improve inference of who-infected-whom. This method will greatly improve understanding of heterogeneity of strain infectivity and help cattle producers to identify highly infectious animals and track transmission routes in farms. It also has the potential to be applied for other important mycobacterial pathogens such as M. bovis.

2.12 MYCOBACTERIUM AVIUM SUBSP. PARATUBERCULOSIS DETECTION IN WILD RABBITS FROM SOUTHERN SPAIN.

Natalia Elguezabal[1], Elena Molina[1], Rakel Arrazuria[1], Iker A Sevilla[1], Mariví Geijo[1], Christian Gortazar[2], Miguel Fuertes[3], Valentín Pérez[3], Ramón Juste[1], Joseba M Garrido[1].

[1]*NEIKER-Instituto Vasco de Investigación y Desarrollo Agrario, Animal Health Department, Derio, Bizkaia, Spain.*
[2]*SaBio, Instituto de Investigación en Recursos Cinegéticos IREC (CSIC-UCLM-JCCM), Ciudad Real, Spain.*
[3]*Department of Animal Health, Faculty of Veterinary Medicine, University of León, Spain*

Mycobacterium avium subsp. paratuberculosis (Map) has previously been detected in wild rabbits in Europe. Studies reporting these findings have focused on rabbits ranging in the proximity of farms with known cases of bovine PTB or in hunting areas inhabited by animals with no information on Map presence.

In the present study, wild rabbits were captured from an area shared with deer affected by PTB with the objective to estimate the prevalence of Map in rabbits in these conditions.

The study sample comprised 33 wild rabbits captured from a hunting estate located in a natural park in Benalup, Cádiz (Spain). Blood was drawn for Map antibody detection by PPA-3 ELISA. At necropsy, carcasses were examined for visible lesions focusing on gut and associated lymphoid tissues. Mesenteric lymph node, vermiform appendix (VA) and sacculus rotundus (SR) were the selected tissues for Map isolation (Herrold´s Egg Yolk agar and 7H10 Egg Yolk agar), mycobacteria detection (triplex PCR for Mycobacterium spp, M. avium and Mycobacterium tuberculosis complex) and histopathological analysis.

Gross lesions consisting in pale white spots in VA and SR were observed in 3 animals (9,1%). Granulomatous lesions compatible with PTB were observed in SR, VA and mesenteric lymph node in 10 of 17 analyzed animals (58,8%). Mycobacterial DNA was detected in 3 animals (9,1%) whereas Map antibodies where detected in 4 animals (12,5%). Cultures were all negative. Altogether, 12 of 33 animals (36,4%) presented a positive result for at least one of the techniques and 6 of 33 animals (18,18%) for two techniques and only one animal for three (3,03%).

These data show higher estimated prevalence by ELISA in wild rabbits of the Iberian Peninsula compared to previous reports. True prevalence may be even higher considering that PTB-compatible lesions were detected in a high proportion of animals and that there were animals positive for more than one method.

2.13 PRACTICAL EXPERIENCES IN DELIVERING A COMMERCIALLY DRIVEN NATIONAL JD PROGRAM IN THE UNITED KINGDOM.

Pete Orpin[1], Richard Sibley[1].

[1]Westridge Veterinary Group

In 2010 Dairy UK (milk processor representative body) formed a Johne's Action Group to tackle JD in the UK dairy sector. The UK JD programme has been built using commercial solutions and synergism's to deliver a National Johne's Management Plan (NJMP) to drive engagement among the farmers, vets and wider industry. Since 2015 the programme has been enhanced by standardised training of vets and processors compelling farmers to engage.

The NJMP aims to ensure that farmers providing 82% of the UK milk supply will by October 2018 will have had a JD risk assessment, appropriate testing and written control plan generated by an accredited NJMP veterinarian.

Phase 1 (2015-17) of the NJMP focused on farmers committing to testing for JD, reviewing their risks and choosing one of 6 control strategies dependent on the risk, prevalence and farmers' aims. The British Cattle Vet Association created an online training program which has delivered 695 accredited JD vets to deliver the program. Phase 2 of the NJMP involves the processors committing their farmers to complete an annual JD risk assessment and written control plan by an accredited JD vet.

Phase 2 of the NJMP will be completed by October 31st 2018. The RESET (Rules, Education, Social Norms, Economics and Tools) model was used to assess the processor and industry views of the programme. The earlier engagement program has delivered the ESET component. However, without the clear synchronised commitment from the prrocessors to deliver the Rules component the NJMP would falter. The JD engagement program had achieved great success with the higher prevalence herds with over 2500 herds committing to strategic milk testing and control. The lower prevalence herds or less engaged farmers require a contractual incentive to comply.

The delivery of a commercially driven national JD program is ambitious. The success depends on creating the appropriate drivers within the RESET behaviour model. The social norm of JD control, risk assessment and management has been created. 8 years after inception of the program the UK programme is entering a crucial phase which aims to deliver widespread of engagement of trained vets and motivated farmers to further reduce the incidence of JD nationally

Poster

2.47 INSIGHTS ON PARATUBERCULOSIS IN CATTLE AT A REGIONAL SCALE USING DATA-DRIVEN MODELING AND INFERENCE

Gael Beaunee[1,3], Pauline Ezanno, Alain Joly[2], Pierre Nicolas[3], Elisabeta Vergu[3].

[1]*INRA - UMR1300 BIOEPAR*
[2]*GDS Bretagne*
[3]*INRA - UR1404 MaIAGE*

Bovine paratuberculosis (agent *Mycobacterium avium* subsp. *paratuberculosis* - Map) is a disease of economic importance whose screening in the field is difficult due to long incubation period and low sensitivity of diagnostic tests. Modern statistics now allow acquiring new knowledge from observed data when combined with mechanistic modelling.

Our objective was to estimate key parameters of a multiscale dynamic model of Map spread from longitudinal and spatial data collected in Brittany (Western France) consisting of serological tests conducted in 2,013 herds sampled between 2005 and 2013. We focused on five parameters: the initial proportion of infected herds and associated within-herd prevalence, the probability of purchasing infected cattle from outside the metapopulation during the considered period, the within-herd transmission rate, and the diagnostic test sensitivity.

Our approach was based on a detailed stochastic model of Map spread within and among dairy herds through animal trad incorporating knowledge of all cattle movements at the regional scale (12,857 dairy herds). Inference relied on a composite-likelihood approximated by Monte-Carlo coupled to an optimization algorithm (Simplex-like). The validity of the inference method was verified on simulated data.

Results confirmed the low diagnostic test sensitivity and provided an average quantitative estimate of 0.28 for this parameter. They also indicated a situation in 2005 with a very large proportion of infected herds (>80%) in the metapopulation but with a low average within-herd prevalence (10%). The estimated proportion of infected animals among those purchased outside and inside the metapopulation were estimated to be roughly similar (~10%), suggesting comparable infection prevalence in different regions.

These estimates of previously unknown parameters provide new insights on Map status in Western France. The inference framework could easily be applied to datasets from other regions concerned by paratuberculosis.

2.48 OCCURRENCE OF SUBCLINICAL PARATUBERCULOSIS IN BUFFALOES FROM BAIXADA MARANHENSE, BRAZIL.

Thais Rocha[1], Emerson Antônio Araújo de Oliveira[1], Isabel Azevedo Carvalho[1], Hamilton Pereira Santos[1], Helder de Moraes Pereira[1]

[1]*Universidade Estadual do Maranhão, Brazil*

The Paratuberculosis is a granulomatous chronic enteritis wich is incurable and it is caused by the *Mycobacterium avium* subspecies p*aratuberculosis* (MAP) that mainly affects ruminants.

The aim of the present study was to *diagnose Mycobacterium avium subsp. paratuberculosis* (MAP) in buffaloes slaughtered in Baixada Maranhense, Maranhão, Brazil.

This is a region formed by low plains that flood in the rainy season, creating huge lagoons between the months of January and June. Samples were collected from small and large intestine, mesenteric lymphnodes and feces from 115 buffaloes (66 females and 49 males), aged above three years, of the Murrah, Mediterranean and mixed races, no clinical signs, in slaughterhouses from Viana and Arari – MA. Histopathological analysis using Hematoxylin and Eosin (HE) and Ziehl-Neelsen (ZN) staining were performed and bacterial isolation was done from fecal samples using Herrold egg yolk medium - HEYM containing mycobactin J.

We observed histopathological alterations suggestive of paratuberculosis in the small intestine, distal portion of the large intestine (colon) and mesenteric lymph nodes in 27% (31/115) of buffaloes slaughtered, and acid fast bacilli (AFB) in 23% (26/115) of them. Five animals that presented lesions of subclinical paratuberculosis when stained by HE did not presented AFB when stained by ZN. MAP was identified in 19% (5/26) fecal samples submitted to bacterial isolation.

The results of this study demonstrate the occurrence of subclinical paratuberculosis in buffaloes slaughtered in Baixada Maranhense and reinforce the necessity of development of an efficient health program to control or eradicate the paratuberculosis in Brazil.

2.49 GENOTYPING OF *MYCOBACTERIUM AVIUM SUBSP. PARATUBERCULOSIS* ISOLATES FROM ARGENTINIAN CATTLE USING MIRU-VNTRS AND SSR (SHORTS SEQUENCE REPEATS).

Roberto Damian Moyano[1]/, Imperiale Belen[1], Romero Magali[2], Alvarado Pinedo Fiorella[2], Santangelo Maria Paz[3], Traveria Gabriel E[2]., Romano Maria Isabel[3].

[1]*IMEX-CONICET Academia Nacional de Medicina*
[2]*CEDIVE- UNLP*
[3]*INTA-Instituto de biotecnología*

Genotyping is a very important tool of epidemiology that helps to trace back the source of infection in case of outbreaks and surveillance programs OBJECTIVE. Simultaneous genotyping of MAP strains using polymerase chain reaction (PCR)-based detection methods: MIRU-VNTR and SSR

Simultaneous genotyping of MAP strains using polymerase chain reaction (PCR)-based detection methods: MIRU-VNTR and SSR.

Isolates of MAP from faeces were achieved on Herrold´s medium with mycobactin J. The IS900-PCR was used to identify MAP and the IS1311-PCR-REA to classify them as C/S type. Genotyping was performed using 8 polymorphic MIRU-VNTR loci (292, X3, 3, 7, 10, 25, 32, 47) and polymorphisms using multilocus short sequence repeats (MLSSR) of four loci (L1, L2, L8 and L9). MIRU-VNTR patterns (INMV) were assigned according with the MAC-INMV database. The strain K10 was used as reference control.

A total of 40 MAP isolates belonging to C type were obtained. These isolates were differentiated into 6 profiles: INMV 1 (18 isolates); INMV 2 (10 isolates); INMV 11 (3 isolates); and the remain isolates were INMV 8; INMV 5; INMV 16. MLSSR revealed loci L1 and L2 as the most polymorphics. SSR locus L1 showed isolates with 7 and 14 repeats and 10 in K10 whereas locus L2 showed isolates with 8, 9, 10, 11, 12, 13 and 17 repeats. Loci L8 and L9 only showed 4 repeats among isolates; only K10 had 5 repeats in those loci. SSR L2 allowed further differentiation of isolates with the same INMV pattern. The 18 MAP isolates with INMV1 were separated in 6 different patterns using SSR L2, each one with different number of G repeats. The 10 MAP isolates with INMV 2, were separated in 4 different genotypes with SSR L2.

We could increase the differentiation among isolates using both genotypic methods. According with our results, MIRU-VNTR assay could be used as a screening tool to differentiate isolates not very close related and then SSR for isolates sharing the same INMV pattern.

2.50 UTILISING FARMER FEEDBACK TO IMPROVE THE UK NATIONAL JOHNE'S MANAGEMENT PLAN

Pete Orpin[1], Dick Sibley[1]

[1]*Westridge Veterinary Practice*

Successful Johne's (JD) control requires a complex mixture of science, policy, practical interventions but most of all an ability to engage and enthuse the farmer to apply the necessary changes at farm level. Dairy UK hosted a conference in February 2017 to launch Phase 2 of the UK National Johne's Management Plan (NJMP). A survey of farmers was undertaken to identify potential challenges and solutions to successful delivery of the NJMP.

A convenience survey was conducted in January 2017 to assess dairy farmers' attitudes to JD control. The results of this survey were shared at the Dairy UK Johne's conference and used to shape the future activities of the NJMP.

A short 8 question survey was circulated to farmers via veterinary practices, milk processors and retailers using the Survey Monkey tool.

394 farmers non-randomly selected completed a survey. 71% had developed a robust JD control plan in conjunction with their vet and a further 22% had created a plan based on their own research and talks. 4% planned to start JD control soon. No farmers surveyed failed to believe in the need to control the risks of JD. 47% of farmers were happy with their control plan and had no concerns. Key concerns cited were insufficient buildings to segregate high risk cattle, confusion with the tests and concerns regarding the costs of culling. The major benefit of effective JD control appeared to be an improved overall health of the herd (83%), reduced forced culling (63%) and improved fertility (50%). Other economic gains were cited such as improved farm margins (40%) and better market opportunities for my processor (42%). Farmers also cited that they had less worry and anxiety now they were in control of JD (41%) The key driver for future engagement was financial incentives from their milk processor (50%), more on farm training on JD practical controls (27%) and more evidence that JD control works (28%) . Further help was requested on understanding risks (24%) and opportunities to visit farms who have controlled disease (20%).

Further farmer engagement with JD control in the UK will be driven by individual milk processor initiatives using a combination of incentives and education to meet the needs of their businesses. Promotion of the wider herd benefits of JD control will be deployed. Veterinary advisor training will be focused providing solutions to the key farmer concerns of testing, segregation and culling for those farmers already engaged and also promoting the benefits of control to those yet to engage with the NJMP. Consistent messaging through www.actionjohnes.org.uk will be an essential tool to support the program

2.51 THE USE OF THE NET PROMOTER SCORE AND FARMER ATTITUDES AND BELIEF TO SHAPE THE UK NATIONAL JOHNES MANAGEMENT PLAN.

Pete Orpin[1], Dick Sibley[1]

[1]*Westridge Veterinary Practice*

Farmer attitudes and beliefs are central to the success of any disease control program. Ritter (2016) divided farmer's attitudes by belief and importance into 4 groups- Proactivists, Unconcerned, Disillusionists and Deniers and highlighted different drivers for engagement existed for each group

The aim was to establish the key drivers required to motivate Unconcerned, Disillusionist and Denier farmers to engage in the UK National Johne's Management Plan (NJMP)

A short 8 question survey was circulated to farmers via veterinary practices, milk processors and retailers using the Survey Monkey tool. The survey was a non random survey.

394 farmers non-randomly selected completed a survey. Of those surveyed, 71% had developed a robust JD control plan in conjunction with their vet and a further 22% has created a plan based on their own research and talks. A further 4% planned to start JD control soon. No farmers surveyed failed to believe in the need to control the risks of JD When asked how they felt about JD control 78% classified themselves as firm believers and would recommend it to other farmers. A further 13% were controlling JD for the benefit of their processor. 7% knew they should control JD but it remained low on their priority right now. The latter two groups were further analysed as an unconcerned group and compared to the Proactivists. Farmers were asked to score themselves on a scale of 1-10 as to on how likely that they would recommend JD control and Protection to another farmer. 51% were classified as Promoters (score 9-10), 28% Passives (score 7-8) and 21% Detractors (score 0-6). The Net Promoter Score (NPS) was 29 indicating that the NJMP was positively accepted. Both Proactivists and Unconcerned farmers supported financial incentives (58% and 47% respectively) and on farm training. However, the unconcerned group required more convincing the rewards outweigh the benefits (31% vs 17% for Proactivists) and more evidence that Johne's control works (49%)

The further success of the UK NJMP will depend on engaging unconcerned farmers. The drivers for engagement of this group will be a mixture of financial incentives, on farm training and additional focus on improving belief in the cost benefit and efficacy of controls. External incentives for engagement will be required for herds of low JD prevalence without a demonstrable within herd cost benefit for control.

2.52 FREQUENCY OF PARATUBERCULOSIS IN GOATS FROM REGIONS III AND IV OF GUANAJUATO, MÉXICO

José Luis Gutiérrez Hernández, Meza Ugalde José María, Herrera López Enrique, Palomares Resendiz Erika Gabriela.

There are no many reports of paratuberculosis (Ptb) in goats from México, It is estimated that the disease has been spreading. Guanajuato is an important state in caprine milk production of México, diseases like Ptb endanger the productive efficiency of the specie.

The aim was to know the frequency of Ptb in goats from III and IV regions of Guanajuato.

910 sanguineous sera from 300 herds was collected by opportunity sampling. Goats >2 years old with corporal condition 1-2 in scale of 5, with Ptb signology were considered. Antibodies against Mycobacterium avium paratuberculosis (Map) were detected by gel- agar immunodiffusion using a protoplasmatic antigen of Map (2mg/ml).

1.2% (11/910) of goats were positive, they are into 8 of 300 herds studied. 5 positive samples belong to III region and 6 to IV region. The herds of III region are intensive and semi-intensive frequently, this characteristic facilitate of spread of Ptb, however, these productive systems have productive and reproductive registers letting the detection and elimination of inefficient, old or sick animals, differing of extensive systems present in IV region. Prevalence of Ptb depends of diverse factors like production system, age of animals, immunological capacity and the sensibility and specificity of the diagnosis tests. The frequency observed in this study could be related with many of these factors.

We conclude that the frequency of Ptb into III and IV regions is low, probably the clinical signs observed are consequence of other affections like parasitosis, physiological state or nutritional deficiencies.

2.53 EPIDEMIOLOGY, 'BIO-TYPE PROFILE' AND BIO-LOAD OF *MYCOBACTERIUM AVIUM SSP. PARATUBERCULOSIS* INFECTION IN DOMESTIC LIVESTOCK, WILD ANIMALS, ENVIRONMENT, MILK, MILK PRODUCTS INCLUDING HUMAN POPULATION AND CONTROL STRATEGIES: INDIAN PERSPECTIVE

Shoor Vir Singh[1], Pravin Kumar Singh[2], Ajay Vir Singh[3], Kuldeep Dhama[4], Krishna Dutta Rawat[5], Sujata Jayaraman[6].

[1]*Animal Health Division, Central Institute for Research on Goats, Makhdoom, PO-Farah- 281 122, Dist. - Mathura, Uttar Pradesh, India.*
[2]*Department of Microbiology, King George's Medical University, Lucknow 226003, India*
[3]*Department of Microbiology & Molecular Biology, National JALMA Institute for Leprosy & Other Mycobacterial Diseases, Taj Ganj, Agra 282001, India*
[4]*Division of Pathology, Indian Veterinary Research Institute, Izatnagar-243 122, Bareilly, Uttar Pradesh, India*
[5]*VTCC, NRC on Equines, HIsar, Haryana, India*
[6]*AIMT & AIB, Amity University Rajasthan, Jaipur, India*

Johne's a major disease endemic in the domestic livestock seriously affect health and productivity.

Estimate bio-load & biotype of *Mycobacterium avium* ssp. *paratuberculosis* in domestic and wild animals, milk, milk products and human beings

In past 32 years (1985-2017), using multiple tests, samples from different sources were screened to estimate bio-load of MAP in the country.

Livestock species-wise bio-load was; Goats-25.1% (5399/21498), Sheep-50.7 (731/1441) {Small ruminants-26.7% (6130/22939)}, Cattle-44.9% (2052/4564), Buffaloes-35.6% (366 /1028) {Large ruminants-43.2% (2418/5592)}.Reported production losses were; Rs.1,840/sheep & Rs. 5,67,176.0/cow in a dairy farm. MAP bio-load was; 48.0% in wild ruminants & other animals (elephant, deer, Bison, blue bulls, moneys, rabbits, hyena, cats, etc.) (196/408); 47.5% in milk & milk products (1106/2329); 64.0% in soil & river water (98/153) & 30.5% in human population (8772/28743). Molecular epidemiology of 555 MAP isolates/DNA from domestic livestock, wild ruminants (blue bulls, bison, deer), wild carnivores, monkeys, human beings and environment, showed 'Indian Bison Type' was highly prevalent (97.1%) followed by 'Cattle type' (2.9%) (2004-2017). Despite failure of 'Test & cull' strategy in goatherds for past 42 years (1976-2017), is still in-use at CIRG, Mathura.

Despite development of highly versatile and robust 'indigenous ELISA kit' for diagnosis of of MAP infection and monitoring of vaccine response, a test to differentiate infected and vaccinated animals (DIVA) and 'therapeutic vaccine' against JD, disease continue to impact productivity and sustainability of domestic livestock. Wherein huge losses continue to occur due to low per animal productivity. In view of the large scale contamination of milk & milk products, there is spurt in human dis-orders of auto-immune nature in the country. Therefore, it is imperative to initiate Nation-wide program on control of JD in domestic livestock to reduce losses & prevent human infection.

2.54 FARMER'S ATTITUDES TOWARDS PARATUBERCULOSIS CONTROL IN CATTLE.

Donat Karsten[2], René Pützschel[1], Veit Zoche-Golob[2].

[1]*Saxon Animal Disease Fund, Animal Health Service*
[2]*Thuringian Animal Disease Fund, Animal Health Service*

Most paratuberculosis control programs worldwide act on a voluntary basis and have to deal with the problem that only a minority of farmers enrolls.

This study aimed at identifying different groups of cattle farmers with similar attitudes towards paratuberculosis control.

In an anonymized survey, questionnaires together with boot-swap sampling sets were distributed among farmers. A total of 255 cattle farmers responded where dairy farmers and large herds were overrepresented. The questionnaire consisted of 27 closed questions about the farmer's opinion and motivations on paratuberculosis control. The farmers' answers were analyzed using a hierarchical cluster analysis. Interpretability of the attitude of the farmers within a cluster and the within cluster homogeneity were prioritized.

Four clusters were identified that represented groups of farmers designated as "sophisticated sceptics", "opposers", "affected supporters" and "free supporters". The sophisticated sceptics did not consider paratuberculosis a dangerous epizootic disease but would enroll in a program if their herd would be affected. The free supporters believed that paratuberculosis is rather dangerous but that their herd is not affected. From these two groups, 80% of the boot swabs were MAP negative. The affected supporters (51.4 % MAP positive) clearly considered paratuberculosis a dangerous epizootic disease. In contrast, the opposers (52.2 % MAP positive) were not concerned about the disease and worried about an image problem of the cattle sector. Generally, the improvement of animal health was among the two most important motivations, and the costs of the laboratory diagnostics were given most frequently as an obstacle. Interestingly, 74.2% of farmers would welcome a mandatory control program in Germany.

In conclusion, the study contributed to a better understanding of the mindset of the decision makers on cattle farms towards paratuberculosis control and showed that different communication strategies are necessary to deal with different attitudes

2.55 EFFECTIVENESS OF ZEOLITE CLINOPTILOLITE AND COPPER OXIDE ON MYCOBACTERIUM AVIUM SUBSP. PARATUBERCULOSIS SURVIVAL

Nicolás González[1], Pamela Steuer[1], Carolina Avilez[1], Carlos Tejeda[1], Alfredo Ramirez-Reveco[1], Miguel Salgado[1].

[1]Universidad Austral de Chile

Mycobacterium avium subsp. *paratuberculosis* (MAP) is the causative agent of paratuberculosis in domestic and wild ruminants, an infectious intestinal disease of great economic importance and related to Crohn's disease in humans MAP is particularly resistant to environmental conditions.

The objective of this study was to evaluate the effectiveness of a novel MAP control system in a liquid medium.

Micronized natural clinoptilolite Zeolite (<10 µm) was used and chemically modified with copper. The study design considered 2 treatments and one control (untreated) with 2 replicates each (Zeolite Clinoptilolite and Zeolite modified with CuO nanoparticles), which were prepared as a filter in lysimeters. Treatment was challenged with highly contaminated PBS buffer (106 MAP CFU/ml). The eluates obtained from each treatment and control were cultured in the BACTEC-MGIT 960 system. In addition, the bacterial cells in the eluate were labeled with fluorescent probes (Life Stain with LIVE / DEAD ™ BacLight ™ Bacterial Viability Kit) in order to determine the viability of the MAP strains.

The samples treated by Zeolite and Zeolite with CuO shown mostly negative cultures and significantly different than controls. Whether this treatment represents an effective and practical decontamination tool for MAP control remains to be determined by further research.

The use of Zeolite modified with CuO demonstrated efficacy in the control of MAP in vitro.

2.56 MOLECULAR CHARACTERIZATION OF BOVINE MYCOBACTERIUM AVIUM SUBSP. PARATUBERCULOSIS FROM KOREAN CATTLE FARMS USING IS1311 PCR-REA, MIRU-VNTR AND MLSSR GENOTYPING

Han Sang Yoo[1], Hong-Tae Park[1], Hyun-Eui Park[1], Woo Bin Park[1], Suji Kim[1].

[1]Department of Infectious Diseases, College of Veterinary Medicine, Seoul National University, Seoul, 08826, Korea

Paratuberculosis (PTB) or Johne's disease (JD) is a chronic and debilitating disease in ruminants, which is caused by *Mycoabcterium avium* subsp. *paratuberculosis* (MAP). Understanding a genetic diversity of MAP isolates is important to the epidemiological studies because it can reveal the infection sources and transmission route and therefore may contribute to the disease control.

The aim of this study was to describe the genetic diversity of MAP obtained from individual cows in South Korea using IS1311 PCR- REA, MIRU-VNTR and MLSSR typing methods.

Twelve MAP positive fecal DNA and 19 MAP isolates were obtained from 10 cattle herds located in 5 provinces in Korea. Additional 5 MAP isolates from Czech Republic and 3 from Australia were used for typing to compare with domestic isolates. 12 Fecal DNA were used only for IS1311 PCR-REA.

Of the 31 Korean fecal DNA/MAP isolates, 8 samples were typed as 'Cattle type', while 23 were typed as 'Bison type' genotypes in IS1311 PCR-REA. All of foreign isolates were shown as 'Cattle type'. MIRU-VNTR and MLSSR typing were conducted using 8 polymorphic loci of tandem repeats (292, X3, 25, 47, 3, 7, 10 and 32) and 11 short sequence repeats. Genotyping using the MIRU- VNTR differentiated 4 subtypes and MLSSR differentiated 7 subtypes among the 27 MAP isolates. The allelic diversity of MIRU- VNTR and MLSSR were calculated as 0.567 and 0.599, respectively. In case of domestic isolates, MIRU-VNTR differentiated only 2 subtypes and MLSSR characterized only 3 subtypes.

In Korea, the 'Bison type' genotype was frequently identified and those were not differentiated by MIRU-VNTR and MLSSR typing. The results indicated that the MAP exhibited low level of genetic diversity across Korea, and additional typing methods are needed to characterize more specifically, especially for 'Bison type'. This work was carried out with the support of "Cooperative Research Program for Agriculture Science & Technology Development (PJ008970)" RDA, BK21 PLUS Program and the Research Institute for Veterinary Science, Republic of Korea.

2.57 MANAGEMENT AND HUSBANDRY PRACTICES ASSOCIATED WITH POTENTIAL TRANSMISSION OF PARATUBERCULOSIS ON DAIRY FARMS

Dick Sibley BVSc HonFRCVS[1], Pete Orpin BVSc MRCVS[1].

[1]*Park Vet Group UK*

Over 90% of dairy farmers using a paratuberculosis management system opted to use risk management to control the disease in their herds. However, considerable practical difficulty in preventing all the potential risks of transmission identified in the risk assessment made total control challenging and jeopardised the outcome of the program.

To identify risks of transmission on dairy farms and highlight potential management practices that may lead to a failure of control

Five dairy farms that had been involved in a paratuberculosis control program but had not achieved expected results were studied to identify where failures in control may have occurred. A comprehensive check list of potential routes of transmission within the herd was used to identify risks that had not been successfully managed.

Numerous risks were identified that had not been robustly managed and could have led to failure in control within the herd. The majority of these uncontrolled risks involved the management of the calving environment and contamination of areas where neonates were kept immediately after birth. The study also identified issues with contamination of milk and colostrum after pasteurisation, and potential transmission from infected or contaminated calves to other calves being kept in the same group. Contaminated feeding utensils, and protective clothing used by workers and vets performing routine tasks on young stock were also identified as significant unconrolled risks.

Robust risk assessment and management plans (RAMP) are a key tool for the control of paratuberculosis. However, their success is dependent upon the implementation of the husbandry practices that potentially spread MAP. This study identified practices that can be easily missed by routine RAMP systems, leading to a failure in control.

2.58 DETECTION OF ANTI-*MYCOBACTERIUM AVIUM SUBSP. PARATUBERCULOSIS* ANTIBODIES IN FREE-RANGING WILD DEER IN THE EASTERN PLAINS AND CARIBBEAN REGION OF COLOMBIA

María de los Ángeles Largo Quintero[1], Jhon Jairo Tuberquia Londoño[1], Mauricio Sánchez Vallejo[1], Mariana Machado Arango[2], Santiago Rodríguez Vélez[2], Santiago Monsalve[2], Nathalia M Correa- Valencia[1], Jorge A Fernández-Silva[1].

[1]*Grupo CENTAURO Escuela de Medicina Veterinaria Facultad de Ciencias Agrarias Universidad de Antioquia*
[2]*Grupo de Investigación en Medicina Veterinaria (GIVET) Corporación Universitaria Lasallista*

Mycobacterium avium subsp. paratuberculosis (MAP) has been reported in farm animals (cattle, goats, sheep, and buffaloes) in Colombia, but neither the agent nor the disease has been reported in wild animal species susceptible to MAP.

This study aimed to determine the presence of anti-MAP antibodies by serum-ELISA in free-ranging deer (Mazama rufina and Odocoileus virginianus) of the Eastern plains and Caribbean region of Colombia.

Specimens were sampled between 2014 and 2016. The ELISA commercial kit Cattletype® MAP (Qiagen) was used to detect the anti- MAP antibodies in serum samples. An animal was considered ELISA-positive at a sample-to-positive ratio (S/P%) of ? 0.4, as recommended by the manufacturer.

Fifty percent (22/44) of animals tested positive, corresponding to 10 females and 12 males. An 36.4% (8/22) and 77.3% (17/22) of these seropositive animals were captured in the Eastern Plains and were reported as adults, respectively.

It is not known how or when MAP was introduced into the Colombian deer population in the regions of study. The most plausible hypothesis to explain the presence of MAP antibodies in these free-ranging deer populations is the transmission by contact from infected cattle widely spread in the regions of study, in which both species shared pastures and commingle. Our findings support the need for further studies evaluating possible links in the infection dynamics between free-ranging deer and domestic animals in Colombia.

2.59 NEW BIOSECURITY AND PARATUBERCULOSIS TOOLS IN QUEENSLAND

Lawrence Gavey[1], Rob Barwell[1].

[1]*Animal Health Australia*

Queensland livestock producers are embracing new arrangements for paratuberculosis management, under broad farm biosecurity planning for animal diseases, plant and crop diseases, weeds, pest animals and chemical contamination risks.

Market incentives drive the new arrangements: the national red meat quality assurance program 'Livestock Production Assurance' (LPA) and interstate movement requirements. Processors demand LPA certification which includes mandatory biosecurity planning from all Australian red meat producers. New industry arrangements for managing paratuberculosis depend on farm level biosecurity measures driven by individual producers' objectives. Australia's new framework no longer supports zoning, state entry requirements or regulatory investigation and tracing. The changes were driven by high costs of regulatory control, decreasing effectiveness associated with producer avoidance, and selective and inflexible standards.

National industry bodies have developed a set of tools to assess, manage and declare paratuberculosis risks. These include guidelines, simple risk scores for beef and dairy, health declarations, and template biosecurity plans. New Queensland legislation specifies a general biosecurity obligation for all people to do what is reasonable and practical to prevent or minimise the likelihood and impacts of biosecurity risks. This law requiring responsible self-management of risks complements the new industry framework.

More than 65 industry workshops have been held, most over-subscribed, with more than 5,500 attendees writing customised biosecurity plans.

Livestock producers in Queensland value the state's historical low prevalence of paratuberculosis, and are acting responsibly to achieve disease management outcomes at least as effective as the former regulatory approach.

2.60 SEROPREVALENCE AND IDENTIFICATION OF *MYCOBACTERIUM AVIUM SUBSP. PARATUBERCULOSIS* IN OVINE FLOCKS OF THE AGUASCALIENTES VALLEY AND ASSOCIATED RISK FACTORS.

Gabriel Ernesto Pallás Guzmán[2], Gilberto Chávez Gris[1], Teódulo Quezada Tristán[2], Edith Maldonado Castro[1], Leticia Chávez González[2].

[1]*Universidad Nacional Autónoma de México.*
[2]*Universidad Autónoma de Aguascalientes*

Paratuberculosis is a disease whose impact on Mexican flocks is unknown

The objective of the study were seroprevalence estimation, MAP identification, and risk determination of factors associated with paratuberculosis.

Obtaining serum samples from 2341 adult animals from 16 herds, the seropositives were identified by a non-commercial ELISA (PPA-3). From samples of ileocecal tissue from 9 animals, bacterial culture was performed in Lowenstein-Jensen medium, and MAP was identified by IS900 PCR. The estimation of the risk factors was obtained from the application of a direct questionnaire with binomial questions and analyzed by a 2x2 contingency table.

Obtaining an apparent seroprevalence of 51.3% and real of 57.7%, with a positive MAP identification in eight of the nine samples. The associated risk factors were, the lack of biosecurity programs with an OR of 4.7, food contamination with feces, feeding in common feeder and not production groups of the flock with an OR of 3.262, and the presentation of nutritional stress and contamination of feeder with feces with an OR of 2.39, the size, the age of the flock and the cleaning frequency had an OR of 1.81.

It is concluded that there is the presence of *Mycobacterium avium subsp. paratuberculosis*, as well as a high prevalence of ovine Paratuberculosis in the Central Valley of the state of Aguascalientes, Mexico. This is strongly associated with the lack of sanitary and biosecurity programs, the contamination with food feces and mangers, the lack of control or subdivision in production and, consequently, poor nutritional management and deficient diets that promote nutritional stress. as the size and age of the herd

2.61 SERO-SURVEILLANCE OF PARATUBERCULOSIS AND BRUCELLOSIS IN DOMESTIC RUMINANT SPECIES OF WESTERN INDIA.

Jagdip Singh Sohal[1], Mukta Jain[1],Rathnagiri Polavarapu[2], B. S. Chandel[3], H. C. Chauhan[3], G. K. Aseri[1].

[1]Amity Institute of Microbial Technology, Amity University RaJasthan, Jaipur, India
[2]Genomix Molecular Diagnostics Pvt. Ltd., Hyderabad, India
[3]College of Veterinary Science & Animal Husbandry, SD Agriculture University, Dantiwada, Gujarat, India

Paratuberculosis and brucellosis are major production diseases of domestic ruminants. Both diseases have zoonotic concerns. Though both diseases are prevalent in India, however, disease surveillance programs are not been in place to know the exact burden of diseases. OIE recommends the serology based screening in routine surveillance programs. In the present study, we did the serological screening of both infections in Western India. Screening of the paratuberculosis was done under the task force project of Indian Council of Medical Research and for brucellosis was done under the network project of Department of Biotechnology, Govt. of India.

To do the sero-surveillance of paratuberculosis and brucellosis in Western India

Serum samples (695) from domestic ruminant species (cattle- 244, buffalo- 78, goat- 266 and sheep- 107) were collected from organized farms, gaushalas (community shelter for let off cattle) and farmer herds from Rajasthan and Gujarat states of Western India. These samples belonged to both suspected and random animals. Samples were subjected commercial ELISA kits for paratuberculosis and brucellosis.

Very high sero-prevalence for paratuberculosis (55.2%) was reported compared to brucellosis (6.9%). Highest positivity was observed in cattle (88.1%) for paratuberculosis followed by sheep (51.4%), buffalo (33.3%) and goat (33%). In case of brucellosis also highest positivity was observed in cattle (13.1%) followed by buffalo (6.41%), sheep (5.6%) and goat (1.8%). In total, 38 (5.4%) and 299 (43.0%) animals were positive and negative for both infections, respectively. Also, 49.9% and 1.5% animals were exclusively positive for paratuberculosis and brucellosis, respectively.

High sero-prevalence of paratuberculosis indicates the wide presence of pathogen in environment resulting in heavy exposure of the domestic ruminants to the pathogen. Low compared prevalence of brucellosis may be indicative of the high slaughter rate of infected/ diseased individuals due to appearance of signs. Signs of the disease in paratuberculosis appear in very late stages so the farmers are unknowing keeping infected animals and as a result high prevalence is reported

2.62 SEROPREVALENCE AND RISK FACTORS ASSOCIATED WITH *MYCOBACTERIUM AVIUM SUBSP. PARATUBERCULOSIS* INFECTION IN SHEEP FLOCKS LOCATED AT SOME REGIONS IN THE DEPARTMENT OF ANTIOQUIA, COLOMBIA

Miguel Hernández-Agudelo[1], Nicolás F. Ramírez-Vásquez[1], Miguel A. Salgado-Alfaro[2], Jorge A. Fernández-Silva[1].

[1]*Escuela de Medicina Veterinaria, Universidad de Antioquia, Colombia.*
[2]*Instituto de Medicina Preventiva Veterinaria, Universidad Austral de Chile, Chile.*

Paratuberculosis causes great economic losses due to a decrease in production parameters and an increase in production cost.

The aim of this study was to determine the presence of antibodies against MAP at animal level and to explore the association between MAP antibodies response and flock level risk factors in some regions of the Department of Antioquia, Colombia.

A total of 456 sheep over 1-year of age in 24 different flocks were studied. In every flock that agrees to participate in the study, 20% of sheep over 1-year of age were sampled randomly, except for flocks with less than 20 adult animals, in which five animals were sampled. Blood was sampled from each animal and serum analyzed using ELISA CATTLETYPE® MAP Ab test (Qiagen, Hilden, Germany). Information on flock management practices was collected using a questionnaire. Bivariable and multivariable logistic regression models were applied to determine the influence of multiple flock management practices and their association with disease seropositivity.

The average population in the flocks was 155 animals. Fifty-four percent of flocks had an area of less than 2 Has, 62% of the flocks had other ruminants (mainly cattle) and in 80% of the cases these species shared paddocks. Fifty-four percent of the flocks shared roads with other farms, 58% purchase sheep and 71% used manure as fertilizer for pastures. According to the ELISA test results 8% (37/456) and 70% (17/24) of sheep and flocks were positives, respectively. The population size was identified as a risk factor for paratuberculosis (p=0,019).

Seropositivity to MAP was detected in a high proportion of the study flocks. Further studies at a regional and national scale considering other diagnostic techniques, such as bacteriological culture and PCR are necessary to determine the distribution of MAP infection in the Colombian sheep industry.

2.63 CO-EXISTENCE OF CATTLE WITH OTHER RUMINANTS IS ASSOCIATED WITH *MYCOBACTERIUM AVIUM SUBSP. PARATUBERCULOSIS* PRESENCE IN ENVIRONMENTAL SAMPLES FROM DAIRY HERDS IN NORTHERN ANTIOQUIA, COLOMBIA

Nathalia M Correa-Valencia[1], Nicolás F Ramírez-Vásquez[1], Jorge A Fernández-Silva[1], Juan C Arango-Sabogal[2], Gilles Fecteau[2].

[1]*Centauro, Escuela de Medicina Veterinaria, Facultad de Ciencias Agrarias, Universidad de Antioquia, Colombia.*
[2]*Département de Sciences Cliniques, Faculté de Médecine Vétérinaire, Université de Montréal, Québec, Canada*

Data on the status *of Mycobacterium avium* subsp *paratuberculosis* (MAP) in Colombian dairy herds remains limited.

This cross-sectional study aimed to determine MAP prevalence based on environmental sampling and to explore herd-level risk factors in selected dairy herds of the Northern region of Antioquia (Colombia), using multivariable analysis.

Study herds (n=292) located in 61 different districts from six municipalities were randomly selected amongst 7,794 dairies registered in the foot-and-mouth disease vaccination records from 2015. The sampling strategy considered optimum and proportional allocation, both at municipality and district level. Herds meeting the inclusion criteria (having adult cattle, in-paddock milking facilities, geographic accessibility, willingness of the owner to participate, and no previous history of MAP detection) were included and visited once between June and October 2016. Each environmental sample contained material from a pool of six different sites of concentration of adult cattle and/or high traffic areas (e.g. areas surrounding waterers and feeders, areas surrounding the current mobile milking-unit place). Identification of MAP was achieved using a duplex Real-time PCR (Bactotype MAP PCR Kit®, Qiagen). A herd was considered as MAP positive if the environmental sample was positive in the Real-time PCR. A risk-assessment questionnaire about general characteristics of the herd, management practices, and knowledge about the disease was administered to the owners/managers. The information on risk factors was analyzed using logistic regression.

The apparent herd-level prevalence was 4.8% (14/292). Risk factors significantly associated with MAP-positive status in the univariable analysis were: co-existence of cattle with other ruminants in the last 2 years, cow manure spreading, and milk-feeding source provided to calves. In the best fit of the model, the co-existence of cattle with other ruminants (OR=3.9; 95%CI: 1.2-13.2) was significantly associated to a positive MAP herd status.

Our results are one of the first approaches towards the definition of MAP herd status in Colombian dairy systems

2.64 FAECAL SHEDDING OF *MYCOBACTERIUM AVIUM SUBSP PARATUBERCULOSIS* IN VACCINATED DAIRY GOAT KIDS

Karianne Lievaart-Peterson[3], Ad Koets[1], Robin Ruuls[2], Maarten F. Weber[3].

[1]*Department of Bacteriology and Epidemiology, Wageningen Bioveterinary Research, Lelystad, The Netherlands; Department of Farm Animal Health, Faculty of Veterinary Me*
[2]*Department of Bacteriology and Epidemiology, Wageningen Bioveterinary Research, Lelystad, The Netherlands*
[3]*GD Animal Health, Deventer, the Netherlands*

Age at onset of faecal shedding of *Mycobacterium avium subsp. paratuberculosis* (Map) is an important parameter in the control of Map. In cattle, transmission of Map amongst young stock has been observed and decreases the beneficial effect of the separation of young stock from adult cattle. However, in dairy goats there is limited information on the age at onset of shedding of Map in faeces.

Therefore, the aim of this study was to quantify the distribution of age at onset of shedding of Map in faeces of dairy goat kids.

Faecal samples were repeatedly collected from a cohort of 151 goat kids from eight dairy goat herds with a confirmed Map infection in the herd. All farmers reported that the goat kids were vaccinated against Map. It was intended to sample each individual five times at 2, 3, 4, 5 and approximately 12 months of age. Faecal samples were tested by IS900 qPCR. The age at onset of shedding was analysed using a Weibull proportional hazards model, taking in to account the asynchronous interval censored nature of the data.

In total, qPCR results of 635 faecal samples from 151 goat kids from 8 herds were available. Map was detected in 43 samples of 38 kids from 7 herds. Of samples collected at 1, 2, 3, 4, 5 to 6 and 11 to 13 month of age, 20%, 7%, 4%, 1%, 0% and 15% were qPCR-positive, respectively ($\chi2=42.7$, $p<0.001$). Samples collected from goat kids of multiparous does were more often qPCR-positive (10%) than samples from goat kids of primiparous does (4%; $\chi2=9.7$, $p=0.002$). The survival analysis confirmed a significant ($p=0.025$) effect of parity of the doe: 16% of kids of nulliparous does were predicted to start shedding before one year of age, in contrast to 38% of kids of multiparous does.

The results indicate that a substantial proportion of dairy goat kids start shedding Map in faeces before adulthood, which raises concerns about the risk of kid-to-kid transmission of Map.

2.65 HERD PREVALENCE ESTIMATES OF MYCOBACTERIUM AVIUM SUBSP. PARATUBERCULOSIS INFECTION IN SASKATCHEWAN, CANADA, BASED ON ENVIRONMENTAL SAMPLES COLLECTED AT TWO SAMPLING PERIODS, 3 YEARS APART

Caroline S. Corbett[1], S. Ali Naqvi[1], Jeroen De Buck[1], Uliana Kanevets[1], John P. Kastelic[1], Herman W. Barkema[1].

[1]*University of Calgary*

Environmental sampling on farm is an effective method for estimating herd prevalence of *Mycobacterium avium subsp. paratuberculosis (MAP)* infection. However, not known is what factors impact the prevalence estimates.

The objectives of the study were to determine: 1) if herd prevalence estimates in the province Saskatchewan, Canada, change when farms are environmentally sampled 3 years following an initial assessment, and 2) if the odds of an environmental sample changing culture status over the 2 sampling times is specific for the location and type of housing (free-stall, tie-stall or loose housing) where the sample has been collected, whether it includes manure of lactating, dry or sick cows (cow group), and whether it is associated with herd size.

From 2012-2013 (sampling 1 (S1)), and 2015-2017 (sampling 2 (S2)), 6 environmental samples were cultured from 148 Saskatchewan dairy farms. Differences between sample period prevalence estimates across different herd sizes and sample characteristics (cow group contributing to sample, type of housing, and location) were estimated using mixed effects logistic regression. Herd-level prevalence was calculated based on sensitivity and specificity estimates of testing MAP-culture positive in at least 1 out of 6 environmental samples.

True prevalence at S1 and S2 was 79 and 48%, respectively. A total of 104 (70%) farms did not change status type from year 1 to year 2. All environmental sample types had decreased MAP-positive results in S2 compared to S1, regardless of sample characteristics types; however, samples collected from the dry cow area had the largest decrease in testing positive in S2. Herds with >200 cows more often tested positive than herds with <51 cows in both S1 and S2.

Prevalence decreased 3 years following the initial assessment, and the change was not specific environmental sample characteristics.

2.66 THE SEROPREVALENCE OF *MYCOBACTERIUM AVIUM SUBSP PARATUBERCULOSIS* IN TWO BEEF CATTLE OF BUENOS AIRES ARGENTINA.

Silvia Colavecchia , Bárbara Fernández, Marcelo Tropeano, Fernando Paolicchi, Silvia Mundo

Paratuberculosis (PTB) is a gastrointestinal contagious chronic disease that affects ruminants caused by *Mycobacterium avium subsp. paratuberculosis* (MAP). Although control programs based on these principles have reduced prevalence of MAP infection in dairy herds, they have generally not eliminated the infection. Prevalence in most regions is currently unknown, In Argentina there are no updated prevalence data, it is estimated between 8 and 20%, and there is no more information because tuberculosis control measures are still being implemented.

The aim of this work is to estimate the seroprevalence of bovine PTB in Argentina.We focused on cattle farms located in Buenos Aires province, which is a major region for beef cattle production

257serum samples from 2 herds of beef farms in Buenos Aires province where PTB has been proven by culture. Sampling was carried out at random and calculating the number of sera with an expected prevalence of 15% and a confidence interval of 95%: Laprida (n=169/4588) and Chacabuco (n=46/59).We developed an in-house ELISA using PPA (Allied monitor) as antigen and anti- IgG bovine HRP (KPL) as second antibody. Positive and negative serum samples were used as controls (Allied monitor).

The seroprevalence obtained were 4.3% (Chacabuco) and 5.9% (Laprida). Animals were cathegorized and percentage of positive samples were calculated: heifers (10.3%), bulls (3.0%), empty cows (5.0%), pregnant cows (5.5%) and none calves were detected as positive. The positive results shows that it may be vary among animal cathegory as was described previously. The percentage of seroprevalence obtained for the breeding herds (between 4.3 and 5.9) shows a difference with the previous data, this could be due to the different management carried out in dairy farms in comparison with the extensive breeding in the province of Buenos Aires.

Then, the seroprevalence here evaluated may change the number of animals to be studied for complete the survey in Argentina.

2.67 EVALUATION OF THE PERFOMANCE OF INSPECTORS IN THE DETECTION PF PARATUBERCULOSIS (OJD) IN SHEEP BY ABATTOIR MONITORING.

Ian J Links[1], Laurence J Denholm[1].

[1]*NSW Department of Industry, Skills and Regional Development, Orange NSW, Australia*

Abattoir monitoring utilising trained inspectors proved a highly sensitive, specific and cost effective procedure for detection of paratuberculosis (OJD) in sheep in NSW Australia from 1999-2009. Assessment of monitoring results for individual inspectors could provide valuable insights into improving the training and work procedures for inspectors.

To review the effectiveness of inspectors in monitoring for OJD in NSW sheep abattoirs from 1999-2009.

From 1999-2009, trained inspectors examined the viscera of sheep in domestic and export abattoirs in NSW for gross lesions suggestive of OJD. The aim was to inspect visually and by palpation the lower small intestine of up to 90% of sheep in each consignment. A maximum of 3 sheep with suspicious lesions (thickened intestinal wall, mesenteric oedema and/or enlarged mesenteric lymph nodes) were sampled for histopathology. Three sites were routinely selected for sampling – terminal ileum, ileocaecal valve and ileocasecal lymph node. The effectiveness of the screening procedure compared with the results from histopathology were evaluated for the individual inspectors.

A total of 32,032 consignments of sheep sourced direct from individual properties were monitored with 5,792 (18.1%) of consignments sampled for histopathology following detection of gross lesions suggestive of OJD. The overall inspection rate for consignments with lesions was 75.2% (2,124,108 killed, 1,597,174 inspected). Monitoring was performed by 36 inspectors in 11 different abattoirs over the 10 year period. Of these, 19 inspectors monitored the majority of consignments sampled (range 22-810) with an average inspection rate of 75.0% (range 54.4%- 95.4%). In the larger export abattoirs sheep were slaughtered at up to 10 sheep per minute, allowing 6 seconds for inspection of the viscera from each animal. The inspectors identified 59,994 (3.8% of sheep inspected) with gross lesions, with 13,101 sheep sampled (av. 2.3/consignment). Histopathology identified 3104 negative and 9,997 OJD positive lesions (1,742 paucibacillary and 8,255 multibacillary). The positive lesions represented 76.3% of all lesions sampled by all 36 inspectors while the 19 major inspectors detected 58,944 lesions, sampled 12,824 sheep and detected 9,763 OJD positive lesions (1,712 paucibacillary and 8,051 multibacillary) representing a positive detection rate of 76.1% (range 42.8% - 96.8%).

This study confirmed the high specificity/low false positive rate of screening (visually and by palpation) of sheep viscera by trained inspectors for lesions attributable to both paucibacillary and multibacillary OJD. They found 76% of samples were attributable to OJD in sheep sourced from all prevalence areas in NSW (combined High, Medium and Low Prevalence). The specificity of screening by individual inspectors in the different prevalence areas and abattoirs over time will be explored further.

Session 3. Pathogenomics and Map Biology

LECTURE: RECENT DEVELOPMENTS IN PATHOGENOMICS TO UNDERSTAND MAP (MYCOBACTERIUM AVIUM SUBSP. PARATUBERCULOSIS) PATHOGENESIS AND DIAGNOSE JOHNE'S DISEASE

DE BUCK, J.
Department of Production Animal Health,
Faculty of Veterinary Medicine, University of Calgary,
Alberta, Canada

The term 'Pathogenomics' groups all comprehensive studies of biological molecules such as proteins, DNA, RNA, small metabolites, lipids or carbohydrates, in relation to disease processes. The integration and merging of different pathogenomic technologies is expected to give the power to resolve complex questions about Map biology and host pathogen interactions in Map infections. Typically, Map pathogenomics are applied to discover biomarkers of infection for diagnostic purposes or in the quest and development of a vaccine strain. Fundamental questions with respect to pathogenesis, evolution driven diversity and host specificity are also within reach but still insufficiently targeted.
In this presentation, an overview will be provided about how pathogenomic technologies are being used in current Map research. The possibilities and opportunities will be laid out on how such questions can be tackled. Recent developments and novel approaches to studying Map biology will be presented.

Genomics: A large number of Map whole genomes are now available from different countries. The phylogenetic relationship of MAP strains and subtypes have been determined and the genetic diversity and distribution the genotypes have been uncovered; however, the evolutionary and fitness consequences of single nucleotide polymorphisms (SNPs) had not been explored. We identified the identify the location and consequences of SNPs in the MAP genomes and determine if particular genes and/or clades showed evidence of differential selection pressures. Separate dN/dS ratios of SNPs present in all isolates and SNPs unique to a single isolate were determined. dN/dS ratios of single isolates and all isolates within a clade were compared. Genes with a higher proportion of nonsynonymous SNPs than expected were identified and subjected to a gene ontology enrichment analysis to identify overrepresented functions. The microevolution of MAP at a limited temporal scale showed a slight trend in the direction of purifying selection over time, although the low number of SNPs decreased the confidence in dN/dS estimates. The identification of genes with an increased proportion of nonsynonymous SNPs provides a foundation to explore these genes and gene functions in more detail.

Tn-seq, a genome-wide technique supported by deep sequencing is a powerful way to identify genes that are important during different stages of the infection. This method can be applied to in vitro and in vivo models alike. Essential genes for survival in an experimental calf infection trial will be presented as well as the follow-up and confirmation studies that are currently taking place in macrophages, mini-gut models, mice, and cattle. Tn-seq applications hold promise to resolve many more aspects of MAP biology and pathogenesis, including MAP's many immune evasion strategies.

Transcriptomics: Microarray studies have revealed some of the host responses during experimental Map infections. A limited number of studies also employed a MAP microarray to study gene expression. RNA-seq, including microRNA-seq, is now increasingly being used to study immune responses against MAP, mostly on in vitro cell culture models, but also in intestinal loop models and in naturally infected cattle. New methods are on the horizon to enable researchers to study the transcripts of host and pathogen. Direct RNA sequencing using Nanopore technology seems very promising in this regard. An overview of previous studies and new opportunities will be presented.

Proteomics: Although not a novel technology, protein crystallography can be very useful in elucidating the function of Map's hypothetical proteins. A selection of essential genes from the Tn-seq screen and Map specific genes from its large sequence polymorphisms are currently being analysed in a high throughput screen to discover new folds and functions. This will lead to new fundamental insights into the pathogenesis of Map.

Metabolomics: The growing field of metabolic profiling (i.e. metabolomics) involves identification and quantification of numerous low molecular weight compounds in biological fluid samples. Metabolomics provides a functional alternative or complement to the above-mentioned techniques, as it measures chemical phenotypes that are the net result of all activity on the transcriptome and proteome levels; therefore, it provides an integrated and reductive profile of the status of the test subject. Previously, a NMR based metabolomic approach was successfully used to study changes in experimentally infected calves. More recently, infected cattle were also studied by direct infusion lipidomics, using electrospray ionization- mass spectrometry. Results from this analysis onindicate that altered availability of choline-containing lipids occurs late in the disease process and is most likely a result of malnutrition and altered biosynthetic capacities of the liver and gastrointestinal tract. Alterations in the bioavailability of these critical structural lipids presumably contributes to the demise of MAP infected cattle. Currently, other metabolomic studies are underway to discover early biomarkers of infection retrospectively on animals that turn positive on common diagnostic tests later on. These studies are comparing different metabolomic technologies, including direct analysis in real time (DART) high resolution mass spec (HMRS) and Headspace–Solid Phase Microextraction–Gas Chromatography (HS-SPME-GC). In this project, age-cohorts from JD-affected farms were studied over a 2-year time period during which each animal were periodically tested for MAP infection by MAP-faecal culture, PCR and ELISA.

Metagenomics: Intestinal inflammation is hallmarks of paratuberculosis. However, there is a paucity of knowledge regarding impacts of the inflammatory responses on the composition and functional properties of intestinal microbiota. So far, only one study has tried to determine the pattern of MAP-associated dysbiosis of intestinal microbiota as a potential biomarker for early detection of infected cattle. Enrichment and underrepresentation of certain metabolomic pathways suggest an interplay of the intestinal microbiota and the immune system. Specific changes in composition of the gut microbiome provided potential biomarker for identifying infected cattle during subclinical stages of JD. Current findings will be highlighted and future opportunities will be presented.

In conclusion, Map pathogenomics research is a quickly evolving and expanding field with great potential to lead to the discovery of biomarkers of Map infection and to clarify host pathogen interactions, resolve immune evasion strategies and hopefully, explain host specificity.

LECTURE: CHRONIC REGIONAL INTESTINAL INFLAMMATORY DISEASE: A WALK THROUGH THE SPECIES

Ramon A. Juste[1], Ana Balseiro[2], Valentin Perez[3]

[1] *Servicio Regional de Investigación y Desarrollo Agroalimentario (SERIDA), 33300 Villaviciosa, Asturias, Spain*
[2] *Servicio Regional de Investigación y Desarrollo Agroalimentario (SERIDA), 33394 Gijón, Asturias, Spain*
[3] *Departamento de Sanidad Animal, Instituto de Ganadería de Montaña (CSIC-ULE), Facultad de Veterinaria, Universidad de León, 24071 León, Spain*

Humans suffer different clinical inflammatory bowel disease (IBD) syndromes of which Crohn´s disease (CD) and ulcerative colitis are the most frequent and best known. IBD has also been reported in non-human primates, dogs, cats and horses. In domestic and wild ruminants this type of chronic granulomatous enteritis is recognized as a pathological condition called paratuberculosis or Johne´s disease caused by the infection with Mycobacterium avium subspecies paratuberculosis (MAP). Due to the similarities between CD and paratuberculosis, MAP has been long considered to be a potential cause of CD, but as yet, evidence to support the role of this, or for that matter, any other specific microorganism in IBD is missing. Based on the clinical course, pathological presentation and inflammatory infiltrate composition, both IBD and paratuberculosis could be included in the term "chronic regional intestinal inflammatory disease (CRIID) syndromes", found in almost any mammal species. A group of infections defined by the mid XXth century as slow infections include paratuberculosis and share a number of characteristics that could help to identify causative elements of the same class.

In particular, all these infections show a defined macrophage component suggestive of phylogenetically primitive innate responses. According to this, CRIIDs may be considered an "intestinal inflammatory slow infection" with a time course varying from weeks to months or even years, and where the inflammatory component would be more prominent in the intestinal interphase between host and environment and be morphologically characterized by an infiltrate ranging from lymphoplasmacytic to histiocytic. Here we review the pathological forms observed in both monogastric and ruminant species CRIIDs in order to identify common characteristics that could lead to a unified pathological classification. From there, a better understanding of disease mechanisms and features throughout different mammal species could eventually point out to similar etiological roots and suggest new strategies for causation demonstration and therapeutic approaches.

Key words: Chronic regional inflammatory intestinal disease; inflammatory bowel disease; Mycobacterium avium subsp. paratuberculosis; histopathological pattern; mammal species

Oral

3.1 ON THE USE OF WHOLE GENOME SEQUENCING TO UNVEIL LOCAL SPREAD OF A PARATUBERCULOSIS CLONE WITHIN A SINGLE HERD.

Matteo Ricchi[1], Simone Russo[1], Erika Scaltriti[2], Simone Leo[1], Luca Bolzoni[2], Norma Arrigoni[1]

[1]Istituto Zooprofilattico Sperimentale della Lombardia e dell'Emilia Romagna, National Reference Centre for paratuberculosis, Strada Faggiola 1, Podenzano (PC), Italy.
Istituto Zooprofilattico Sperimentale della Lombardia e dell'Emilia Romagna, Risk Analysis Unit, Via dei Mercati 13/A, Parma 43126, Italy.

Mycobacterium avium subsp. *paratuberculosis* (MAP) is characterized by a low genomic rate of mutation. Current subtyping tools, such as Mini-Micro-satellite analyses, have not sufficient discriminatory power to follow MAP's evolution on a small spatial and temporal scale.

Aim of the study was to investigate the evolution and spread of MAP within a single dairy herd through whole genome sequencing (WGS).

For this purpose, the DNAs of 48 field isolates (with identical satellite subtype), collected in the 2012-2016 period from faeces of cows of the same herd, were sequenced by Miseq system (Illumina). The herd rarely bought new animals and in one case, it was possible to identify an isolate collected from a cow born outside the herd (bread Ottonese).

The comparison of the genomes found no indels and 44 polymorphic sites, seven of which were shared between two or more isolates. Bayesian evolutionary analysis showed a mutation rate compatible with that previously reported. The MAP population evolves according to a strict clock model, confirming the clonal origin of the infection. Moreover, according to herd prevalence data, the coalescent model assuming constant population provided a better fit of the clone evolution than that assuming a growing population. Four separate MAP sub-lineages were identified within the herd, suggesting a sufficient phylogenetic signal to investigate transmission patterns. The observed patterns are consistent with the life history data since the ancestral sub-lineage displays larger genetic variability and it was found in the older cows. The other lineages emerged concurrently in more recent years indicating heterogeneity in within-herd transmission. The isolate collected from the Ottonese cow clustered with the others suggesting the infection occurred after the introduction in the herd.

These findings show that WGS, coupled with appropriate epidemiological information, represents a valuable tool to provide new insight on MAP's spread and evolution at herd level.

3.2 EVALUATION OF A MAP MUTANT IN BOVINE BMDM AND IN A MOUSE MODEL, AND NEW METHODS FOR M. AVIUM SUBSPECIE PARATUBERCULOSIS MUTAGENESIS.

María de la Paz Santangelo[1], Mariana Viale[1], Alejandra Colombatti[1], Natalia Alonso[1], William Davis[2], María Isabel Romano[1].

[1]*Instituto de Biotecnología, INTA. Buenos Aires, Argentina. CONICET*

[2]*Department of Veterinary Microbiology and Pathology, College of Veterinary Medicine, Washington State University, Pullman, WA 99164, USA*

Several strategies have been used to obtain mutants of *Mycobacterium avium* subsp. *paratuberculosis (*Map). Transposon mutagenesis was applied to construct various libraries. A specialized transduction system originally developed for M. tuberculosis was adapted to improve the efficiency of allelic exchange in order to generate site-directed mutations in preselected genes in Map. Recent advances have been made in a new strategy to produce phages using recombineering methods. This method was successfully developed for efficient genetic manipulation of bacteria. We are adapting this method to obtain mutants in the mce4 operons of Map. This operon may facilitate uptake and utilization of cholesterol from host cells during infection, a function which is known to be essential for successful cellular invasion and persistence in Mycobacterium spp.

Using the specialized transduction system we obtained a Map lprG mutant (DlprG). LprG is a cell envelope lipoprotein, required for LAM translocation to the cell surface known to be an important virulence factor in M. tuberculosis and M. bovis.

We studied the effect of DlprG in bovine BMDM entry, intracellular survival and we evaluated its replication in spleen of infected mice.

Initial screening in bovine BMDM showed that the mutation of lprG impairs Map replication inside bovine macrophages compared to wild-type MAP K10 at 2, 4, 6 and 8 days post infection. These results are in agreement with previous studies in M. tuberculosis and M. bovis. However, as ex vivo models do not usually predict survival in vivo, BALB/c mice were infected with ΔlprG. Although this experiment is ongoing the results indicate the clearance of the mutant in the spleen as soon as 3dpi.

These data will allow researchers in the field to determine the best strategy to obtain mutants in Map in order to study gene functions. This study also underlies the development of modified live vaccines, one of the best strategies for controlling Johne's disease.

3.3 MYCOBACTERIUM AVIUM SUBSP. PARATUBERCULOSIS: IN VITRO INTERACTION WITH AND BOVINE SPERM CELLS.

Isis Espeschit[1], Ana Carolina Silva Faria[1], Daniel Mendonça de Araújo Lima[1], Breno Soares Camilo[1], David Germano Gonçalves Schwars[1], Mariana Barros[1], Sanely Lourenço da Costa[1], Maria Aparecida Scatamburlo Moreira[1].

[1]*Universidade Federal de Viçosa.*

The economic impact created by paratuberculosis occurs mainly in dairy herds, relating to the decrease of milk production, low reproductive efficacy, premature slaughter and reduction of carcass value. *Mycobacterium avium* subsp. *paratuberculosis* (MAP) has already been isolated from semen and reproductive organs of infected bulls and rams, also, there is evidence that it can survive semen conservation procedures containing antibiotic in liquid nitrogen.

The aim of the present study was studying the interaction between MAP and frozen bovine sperm cells.

A straw of frozen semen was thawed in a 37 °C water bath for 30 s prior to the removal of the medium by centrifugation. The sperm cells were washed and transferred to a 199 culture medium supplemented with 10% fetal bovine serum, MAP K-10 strain was added in a cell ratio of 1:1 (bacteria:sperm cells). This mixture (bacteria-sperm cells) was maintained at 37 ºC for 1, 4 and 6 hours. At each moment, this mixture was fixed and multiple fields were analyzed by Scanning Electron Microscopy. MAP k10 and sperm cells alone were used as controls.

At all analyzed intervals, MAP attached on spermatozoa only at tail midpiece. A greater amount of MAP attached in tail midpiece was observed at 4 and 6 hours, compared with previous intervals.

The midpiece of the spermatozoa is a fibronectin rich region that has a central filamentous nucleus with many ATP producing mitochondria, energy used by the sperm to travel through the cervix, uterus and uterine tubes until it reaches the ovum. In this way, MAP presented in this part may alter this journey, resulting in problems in fertilization since it is known that the midpiece can also enter the ovule. Further studies should be performed to investigate the motif of this MAP localization in bovine sperm and the actual resulting reproductive problems.

3.4 ASSESSMENT OF MYCOBACTERIUM AVIUM SUBSP. PARATUBERCULOSIS VIABILITY USING FLOW CYTOMETRY.

Timothy Scott[1], Marina Buciuc[1].

[1]Minnesota State University, Mankato

Assessment of the viability of slow-growing microorganisms has always been laborious and time-consuming. Viability assessment kits using fluorescent dyes enable actively growing cells to be distinguished from those that are senescent or dead based on cell membrane integrity. SYTO9 is a membrane-permeant dye that stains all cells regardless of cell membrane integrity, whereas propidium iodide (PI) penetrates cells lacking an intact membrane. Unfortunately, the use of these assays with *Mycobacterium avium* subsp. *paratuberculosis* (MAP) often fail to yield expected results when the supplier's recommendations are followed.

We modified the protocols for two commercially-available viability kits to evaluate their ability to assess viability of mid-to-late log phase MAP cultures.

Suspensions of live or killed mid-to-late log phase MAP cells were prepared by incubating cells in 70% isopropyl alcohol or sterile phosphate-buffered saline (PBS), respectively, after which cells were washed and resuspended in PBS. Live and dead cells were mixed to generate populations of cells ranging from 100% to 0% viable cells and combined with differing concentrations of SYTO9 and PI dissolved in either dimethyl sulfoxide (DMSO) or water. The samples were read using a flow cytometer equipped with a 488nm laser.

Fluorochromes dissolved in water failed to facilitate the accurate sorting live cells from those that were dead or dying. However, flow cytometry yielded two distinct populations of cells which stained intensely with either SYTO9 or PI when the two dyes were suspended in DMSO and added to MAP in equal proportion. Analysis of the data generated by flow cytometry yielded results comparable to those expected for exponentially-growing cells.

This protocol enables the rapid, and accurate enumeration of viable MAP in an experimental setting.

3.5 TRANSCRIPTOME ANALYSIS OF ILEUM TISSUES FROM MYCOBACTERIUM AVIUM SSP. PARATUBERCULOSIS INFECTED COWS REVEALS IMPORTANT GENES AND PATHWAYS FOR JOHNE'S DISEASE.

Nathalie Bissonnette[1], Duy N. Do[1], Pier-Luc Dudemaine[2], Eveline M. Ibeagha-Awemu[1].

[1]Agriculture and Agri-Food Canada, Sherbrooke Research and Development Centre, Sherbrooke, Qc, Canada

[2]Agriculture and Agri-Food Canada, Sherbrooke Research and Development Centre, Sherbrooke, Department of Animal Science, McGill University, Qc, Canada

Mycobacterium avium ssp. *paratuberculosis* (MAP) infects ruminants, reside in the intestinal submucosa, invades and slowly proliferates in macrophage/dendritic cells. The pathology associated with tissue change due to extensive formation of granulomas causing Johne's disease (JD) is poorly understood.

This study aimed to explore the transcriptomic modifications in ileum tissues of cows naturally infected with MAP.

Total RNA from ileum tissues collected from five JD positive and five negative cows were subjected to RNA sequencing, bioinformatics and pathways analyses.

About 260 genes were significantly differentially expressed (DE) between negative and JD cows (pFDR<0.05). The most significant DE genes were ENSBTAG00000048135 (pFDR=1.25E-19), ACADS (pFDR=1.41E-08) and KLF10 (pFDR=6.17E-08). ENSBTAG00000048135 is orthologue with human IGHG4 gene known to play roles in phagocytosis and innate immunity. KLF10 is an important transcription factor involved in mediating TGF-β1 signaling. DE genes were significantly enriched in many biological processes gene ontology (GO) terms involved in immune regulation (neutrophil degranulation, acute inflammatory response and positive regulation of cytokine production), thus stressing the importance of immune processes for JD. The most enriched cell components GO terms were transmembrane collagen trimer, IL-20 receptor complex and NarGHI complex. Toll-like receptor (TLR) signaling pathway, known as important for cellular response to MAP infection, was the most significantly enriched pathway. Lysosome was the second most enriched pathway indicating that DE genes might contribute to MAP infection via different processes (endocytosis, phagocytosis, and autophagy) within this pathway.

In summary, this study provides insight on the genes and pathways related to MAP infection and confirmed the importance of genes in TLR and lysosome pathways in the regulation of ileum functions during MAP infection.

3.6 HOST GENE EXPRESSION SIGNATURE OF IMMUNE-REGULATORY GENES IN WHOLE BLOOD OF CATTLE WITH SUBCLINICAL INFECTION OF MYCOBACTERIUM AVIUM SUBSP. PARATUBERCULOSIS.

Han Sang Yoo[1], Hyun-Eui Park[1], Hongtae Park[1], Young Hoon Jung[2].

[1]Department of Infectious Diseases, College of Veterinary Medicine, Seoul National University, Korea.

[2]Department of Animal Resources Development, National Insitute of Animal Science, Rural Development Administyration, Korea

Paratuberculosis (PTB), or Johne's disease (JD), is a chronic wasting disease of ruminants caused by infection with Mycobacterium avium subsp. paratuberculosis (MAP). JD is characterized with persistent diarrhea and cachexia after the long subclinical period. Host response against MAP infection in early stage mainly regulated by Th1 type immune response which characterized with production of IFN-γ. With the progression of disease, MAP can survive and proliferate in the host through the evasion of host immune response by manipulating host immune response.

However, the host response induced by infection of MAP during subclinical phases has not been fully explored. Immune regulatory genes, including Th17-derived cytokines, interferon regulatory factors, and calcium signaling-associated genes, are hypothesized to play a crucial role during subclinical phases of Johne's disease.

Therefore, the present study was tried to analyze the host gene expression signatures of immune regulatory genes during MAP infection in whole blood. Distinct host gene expression signatures were identified depending on the different level of ELISA S/P ratio and presence of fecal shedding.

Decreased expression of IL17A, IL17F, IL22, IL26, HMGB1, and IRF4 and increased expression of PIP5K1C indicate suppression of the Th1 response and loss of granuloma integrity which can cause fecal shedding. In addition, up-regulation of IRF5 and IRF7 suggest activation of IFN-α/β signaling during subclinical stages, which can induce inhibition of T cell proliferation. Up-regulation of CORO1A indicate modulation of calcium signaling, which enhanced the survival of MAP.

Taken together, distinct host gene expression signature induced by MAP infection suggests enhanced survival of MAP during subclinical stages. This study was supported by the Cooperative Research Program for Agriculture Science & Technology Development (Project No. PJ00897001), Rural Development Administration, BK21 PLUS and the Research Institute for Veterinary Sciences, Seoul National University, Republic of Korea.

Poster

3.68 STUDIES ON THE 'HOST BIOMARKERS' FOR THE GENETIC SUSCEPTIBILITY TO MYCOBACTERIUM AVIUM 2,PARATUBERCULOSIS INFECTION IN HUMAN PATIENTS SUFFERING FROM ABDOMINAL DISORDERS IN INDIA, USING REAL-TIME BASED REVERSE TRANSCRIPTION PCR

Saurabh Gupta[1], Shoor Vir Singh[1], Manju Singh[1], Kundan Kumar Chaubey[1], Abdul Qayoom MirF[2], Adil Majid Bhat[3].

[1]Animal Health Division, ICAR-Central Institute for Research on Goats, Makhdoom, PO-Farah, Mathura - 281122 (UP), India
[2]Faculty of Veterinary Sciences, Sher-e-Kashmir University of Agricultural Sciences and Technology, Shuhama, Jammu & Kashmir, India.
[3]Division of Veterinary Clinical Medicine, FVSc & AH, Shuhama, SKUAST-K, Srinagar, Jammu & Kashmir, India

Johne's, a major disease endemic in the livestock population of the country, is a notifiable disease seriously affecting health and productivity. Screening of samples from different sources from 1985 to 2017, using multiple tests, the bio-load of MAP was 29.9% in domestic livestock species; 48.0% in wild ruminants; 47.5% in milk & milk products and 30.5% in human population.

We aimed to study host gene expression during *M. paratuberculosis* infection in chronic abdominal disorders.

Using microscopy, IS900 PCR and indigenous ELISA kit, we diagnosed 32 patients suffering from abdominal disorders as infected with Mycobacterium avium ssp., paratuberculosis (MAP) and 5 of them responded to anti-MAP treatment. In another study we estimated bio-load of MAP in human population suffering with symptoms of Crohn's disease in North India. Patients infected with MAP had predominance of gene-variants involved in susceptibility to MAP in Crohn's disease patients. Real-time based reverse transcription PCR was optimized and standardized for three susceptibility gene candidates (TLR2, TLR4 and NOD2) to diagnose MAP infection.

Expression of TLR2, TLR4 and NOD2 genes was calculated by ratio to housekeeping gene (human Ribosomal Protein Large P0-RPLP0), which was amplified in parallel reactions. All three genes (TLR2, TLR4 and NOD2) were up-regulated in MAP-infected humans patients suffering with Crohn’s disease ($p < 0.05$).

The study suggested genetic susceptibility to MAP and its clinical progression in patients suffering with Crohn's disease.

3.69 PATHO-MOLECULAR STUDIES ON PARATUBERCULOSIS IN BUFFALOES (BUBALUS BUBALIS) IN MALWA REGION OF MADHYA PRADESH

Gayaprasad Jatav[1], U. K. Garg[1], Shoorvir Singh[2], Ajay Gautam[3], Shalini Sharma[3], Naveen Kumar[4].

[1]*Department of Veterinary Pathology, College of Veterinary Science and Animal Husbandary, Mhow, MP, India.*
[2]*Animal Health Division, ICAR-Central Institute for Research on Goats, Makhdoom, PO-Farah, Mathura - 281122 (UP), India.*
[3]*Department of Veterinary Physiology & Biochemistry, Lala Lajpat Rai University of Veterinary & Animal Sciences, Hisar-125001, Haryana, India.*
[4]*National Center for Veterinary Type Cultures, ICAR-NRC on Equines, Sirsa Road, Hisar- 125001, Haryana, India.*

Water buffalo (Bubalus bubalis), at 110.0 million heads (2016), is important domestic animal for milk, meat and draught puroposes in India ('Black diamond').

We aimed to study pathomorphology and patho-molecular profile of the lesions in subclinical versus clinical Johne's disease.

Efficacy of histopathology in conjunction with acid fast staining was compared with ELISA and IS900 blood PCR and other diagnostic tests for the diagnosis of JD in 150 buffaloes ('off-production') slaughtered for harvesting meat in Malwa region in Madhya Pradesh.

On the basis of gross lesions, incidence of MAP infection was 87.3% (18.3 sub-clinical; 81.6% clinical lesions). On physical profile basis, sub-clinical infection was recorded in 6.0% buffaloes. Faecal, intestinal and mesenteric lymph nodes (MLN) smears revealed, 79.3, 64.6 and 49.3% infection, respectively. Grossly tissues were thickened, congested, haemorrhagic, oedematous mucosal folds of intestine and on ileo-caecal junction mucosa along with lymphangitis. Adjacent MLNs were enlarged, oedematous and congested in medullary region. Histopathology of tissues revealed congestion, oedema, infiltration of inflammatory cells (plasma cells, epithelioid cells, lymphocytes, macrophages and eosinophils) in H & E and ZN staining of intestine, lamina propria and sub mucosa as well as on epithelium of villi and in sub-cortical, subcapsular, medullary sinuses and medullary cords of associated MLNs. Granulomatous inflammation and granuloma were evident in dorsal portion of villi, submucosda of intestine and in cortical area MLNs. Indigenous ELISA kit, dot-ELISA, Latex agglutination test and fluorescent antibody test revealed, 89.3, 82.1, 52.6 and 55.0%, respectively were positive for MAP infection.

Study reported high bioload of MAP in slaughtered buffaloes and bio-typing by IS1311 PCR-REA revealed 'Indian Bison Type' bio- type infecting buffaloes of Malwa region of Madhya Pradesh.

3.70 MAP3773C A ZUR PROTEIN: BIOCHIMICAL ANALYSIS AND REGULATORY FUNTION IN THE ZINC REGULON OF MYCOBACTERIUM AVIUM SUBSP. PARATUBERCULOSIS

Bertha Landeros Sanchez[1], José A. Gutiérrez – Pabello[2], German Ibarra Molina[1], Yunuen Nieto Morin[1], Alexes González Cornejo[1], Lilia A. Hurtado Ayala[1], Mirna C. Brito Perea[3], Esteban Hernández Guevara[3], Dania M. Díaz Padilla[3], Ángela Fernández Otal[3], María Fillat[3].

[1] *Universidad Autónoma de Baja California, Facultad de Ciencias Químicas e Ingeniería.*
[2] *Facultad de Medicina Veterinaria y Zootecnia, Universidad Nacional Autónoma de México.*
[3]*Instituto Aragonés de Ciencias de la Salud and Department of Biochemistry and Molecular and Cell Biology. Faculty of Sciences. University of Zaragoza.*

Paratuberculosis (PTB) is caused by *Mycobacterium avium* subsp. *paratuberculosis (*MAP).

This study has designed to characterize the regulatory functionality and biochemical characterization of the Zur MAP3773c protein.

MAP3773S was induced with IPTG, Fe+++, Zn++ and NTA. We performed the purification protein induced IPTG in the IMAC-Zn++ column. We determine structural content of Zinc with PAR staining, gluraldehyde crosslinking, molecular mass and its self-regulation and Zinc regulon regulation by EMSA, and the partial secondary structure by circular dichroism using DicroWeb, was determined.

The MAP3773c protein has not prosthetic group, does not form intramolecular binding oligomers by DTT analysis, does not form oligomers with glutaraldehyde, and has structural Zn. The alpha helix is present in 50% folded beta sheet 20% also in 7% of the turns and in 23% of structurally disordered sequences. MAP3773c was expressed with IPTG, Fe+++, Zn ++ and NTA. MAP3773c is self- regulated and regulated the map3736c, map3737 and map3739c of Zinc regulon.

MAP3773c is a protein constitutive and it is able to bind to two iron boxes of bacteria such as E. coli and the mycobacterial box found in the promoter of its gene. That the MAP3773c protein is involved in its self-regulation and Zinc regulon.

3.71 DECIPHERING THE VIRULENCE AND INTERACTION OF MYCOBACTERIUM AVIUM SUBSP. PARATUBERCULOSIS (MAP) ISOLATES WITH ANIMAL HOST MACROPHAGES USING MATHEMATICAL MODELS

Marta Alonso-Hearn[2], Gesham Magombedze[1], Naiara Abendaño[2], Mariana Landin[3], Ramon A. Juste[4].

[1]*Center for Infectious Diseases Research and Experimental Therapeutics, Baylor University Medical Center, Dallas, TX, USA.*
[2]*NEIKER-Instituto Vasco de Investigación y Desarrollo Agrario, Animal Health Department, Derio, Bizkaia, Spain.*
[3]*Department of Pharmacology, Pharmacy and Pharmaceutical Technology, Universidad de Santiago de Compostela, Santiago de Compostela, Spain.*
[4]*Agri-Food Research and Development Regional Service, SERIDA, Villaviciosa, Asturias, Spain*

Understanding why some pathogenic Map isolates cause disparate disease outcomes with differing magnitudes of severity is important in designing new control strategies.

We applied a suite of mathematical models: i) general linear, ii) neurofuzzy logic, and iii) a logistic growth ordinary differential equation (ODE) to explain how the host of origin of several Map isolates (cattle, sheep, goat, red deer, fallow deer, wild boar, or bison), Map genotype (C, S or B), host (bovine and ovine), macrophage-based in vitro models (BOMAC, ovine MDM, bovine MDM) and time post-infection (2 h and 7 days p.i.) contributed (alone or combined) to the infection.

The bacterial growth ODE model was also applied to estimate within macrophage growth rates for the different Map isolates. In addition, we correlated bacterial growth within infected macrophages with the levels of several expressed innate immune markers for some representative Map isolates.

The models predicted different susceptibilities of bovine and ovine macrophages to Map infection and confirmed distinct virulence profiles for the isolates, judged by their ability to grow within macrophages. Estimation of Map replication within macrophages revealed that the Map isolates from sheep (P381, 2349) are inclined to persist within macrophages through the activation of anti-destructive responses. However, growth of the bovine K10 strain within macrophages was shown to be negatively correlated with the expression of the apoptotic inducer TNF-α. The ability of the bovine isolates (K10 and 6) and the non-domesticated animal isolates (622, 681, 6.1) to grow with higher CFU numbers within bovine macrophages suggest that these isolates are more virulent than the sheep and goat isolates.

Overall, our results confirm the different virulence levels for the Map isolates and susceptibility profiles of host macrophages, which is crucial in increasing our understanding of Map infection.

3.72 NATURAL TRANSMISSION OF MYCOBACTERIUM AVIUM SUBSP. PARATUBERCULOSIS FROM A MIXED BREED BOER GOAT TO A VIETNAMESE POT-BELLIED PIG.

Lucy R. Hancock[1], Christine Ward[1], Jaclyn Dykstra[1], Janes E. Collins[1], Adel Talaat[2], Michael T. Collins[2].

[1]*University of Minnesota, College of Veterinary Medicine.*
[2]*University of Wisconsin, School of Veterinary Medicine.*

A 12-year-old male neutered Vietnamese pot-bellied pig developed progressive inappetence, lethargy and diarrhea. The physical exam showed normal body condition, normal gastrointestinal sounds, and no tenderness on abdominal palpation.

A complete blood count showed a marked increase in neutrophils (34.4 x 103/uL). Serum chemistries indicated normal kidney and liver function. However, the pig had low total serum protein (TP) and albumin (A) values (TP 4.1 g/dL; A 1.5 g/dL). Abdominal ultrasonography revealed a prominent small intestine with thickened walls throughout the abdomen. Intestinal loops were filled but not distended and motility was reduced.

This pet pig was a long-time companion to an adopted mixed breed Boer goat that died from paratuberculosis two and a half years prior to onset of the pig's clinical signs (confirmed by histopathology and PCR). For this reason, and because of the ultrasound and serum chemistry findings, a fecal sample was submitted for Mycobacterium avium subsp. paratuberculosis (MAP) IS900 PCR.

The PCR was positive with a Ct value of 20.8 (cut-off for a positive = 35) indicating the pig was a heavy shedder of MAP. The pig's condition continued to decline despite therapy, and was euthanized. Necropsy and histopathology confirmed a diagnosis of paratuberculosis. Cultures recovered MAP. Whole genome sequencing is in progress to compare MAP strains isolated from the goat and pig.

This represents the first report of paratuberculosis in a pot-bellied pig and the one of three reports of natural transmission of paratuberculosis to swine. This finding verifies that swine, in particular pot-bellied pigs, are susceptible to MAP and potentially provides a useful "small" animal model for study of a Crohn's disease-like illness.

3.73 LIVER TRANSCRIPTOME AND NETWORK ANALYSES REVEAL GENES AND BIOLOGICAL PROCESSES AFFECTED BY MYCOBACTERIUM AVIUM SSP. PARATUBERCULOSIS INFECTION

Nathalie Bissonnette[1], Duy N. Do[1], Pier-Luc Dudemaine[1], Eveline Ibeagha-Awemu[1].

[1]*Agriculture and Agri-Food Canada, Sherbrooke Research and Development Centre, Sherbrooke AND Department of Animal Science, McGill University, Quebec, Canada.*

Mycobacterium avium ssp. paratuberculosis (MAP) causes Johne's disease (JD) in cows. JD progression involves a general deterioration of health, progressive weight loss, general wasting, and decreased milk production. Liver is an important organ for metabolic activities and its functioning might be altered by JD.

This study aimed to explore the major liver genes and biological processes affected by MAP infection.

Total RNA from liver tissues of three JD positive and five negative cows were subjected to mRNA sequencing, bioinformatics and pathways analyses.

A total of 15 and 1130 genes were significantly differentially expressed (DE) between JD positive and negative cows at pFDR<0.05 and uncorrected p<0.05, respectively. A novel gene (ENSBTAG00000032642) was the most DE (pFDR=3E-06) while other genes (LRP2, LPIN1 and STRIP2) with known functions were also highly DE (pFDR<5E-04). 29 KEGG pathways were enriched (p<0.05) for DE genes. The most enriched pathways were adherent junction, complement and coagulation cascades, and ribosome as well as disease and immune related pathways (e.g. Salmonella infection, Staphylococcus aureus infection and cytokine-cytokine receptor interaction). Using DE genes (n=1130), weighted gene co-expression networks were constructed. Five co-expressed gene modules (groups of genes with similar expression patterns) named BLUE (664 genes), YELLOW (95 genes), BROWN (266 genes), PINK (51 genes) and MAGENTA (46 genes) were defined, and LPIN1, LSM8, HNRNPAB, SLC9A9 and LRRC56 were hub genes for these modules, respectively. The most enriched (P<0.05) pathways for module genes were metabolic pathways (YELLOW and PINK), adherent junction (BLUE) and ribosome (BROWN and PINK).

Overall, transcriptome analysis suggests that MAP infection affected gene expression and hepatic metabolic pathways in the liver of JD positive cows. This study thus provides an insight into the indirect effect of MAP infection on an important metabolic controlling organ.

Session 4. Public Health and Map Environment.

LECTURE: CROHN'S DISEASE AND JOHNE'S DISEASE: WHY DOESN'T EVERYONE BELIEVE YOU?

RODRICK CHIODINI

Since the mid-1980's, there has been increasing evidence of an association between Crohn's disease and *M. paratuberculosis*. However, convincing evidence of causation, as opposed to simple association, has not been forthcoming and the scientific community has general not been supportive. This lack of support stems from the alienation and skepticism of groups essential to the overall acceptance of a link between *M. paratuberculosis* and Crohn's disease. In this presentation, the issues that have alienated and caused skepticism among scientific groups will be discussed including leaps of faith in the interpretation of data, obvious and never-ending bias in publications and other avenues, the redefinition of terms to support positions that end up discrediting other research data, the lack of focus on a single disease rather than a host of other idiopathic diseases, erroneous extrapolations and comparisons to other diseases, embellishment of data, unexplained and conflicting data from reputable scientists, conflicts of interest, and the unwillingness to consider constructive criticisms. All these factors contribute to the discount and the lack of acceptance of an etiologic role of *M. paratuberculosis* in Crohn's disease.

Oral

4.1 NOVEL NANO-IMMUNO-TEST FOR THE DETECTION OF LIVE PARATB BACILLI IN THE MILK OF DOMESTIC LIVESTOCK.

Shoor Vir Singh[1], Manju Singh[1], Bjorn John Stephan[2], Manali Dutta[2], Gajendra Kumar Aseri[2], Jagdip Singh Sohal[2].

[1]*Animal Health Division, Central Institute for Research on Goats, Makhdoom, PO-Farah- 281 122, Dist. - Mathura, Uttar Pradesh, India.*
[2]*AIMT & AIB, Amity University Rajasthan, Jaipur, India.*

Early rapid detection of *Mycobacterium avium* subsp. *paratuberculosis* in milk is major challenge since present assays are time consuming, laboratory dependent and cannot be performed in the field. We report a simple, sensitive and specific nano-technology based (Nano-immuno test) that can detect viable MAP bacilli in milk samples in 10 hours.

To detect viable MAP bacilli in milk samples using Nano-immuno diagnostic test

Test was optimized using true positive (10-bovine, 12-goats) and true negative (16-bovine, 25-goats) culture positive raw milk samples collected from domestic livestock endemically infected with MAP. MAP bacilli in milk samples were captured by MAP specific antibody-conjugated magnetic nano-particles. MAP-MAb-MNP complexes separated application of external magnetic field and numeration of live bacteria was mediated by resazurin dye. This novel nanotechnology based chromogenic method enabled detection of viable MAP present in the milk samples.

Goat milk samples (37) were screened by this test. Of 12 (32.4%) culture positive milk samples, 9 (24.3%) and 11 (29.7%) were positive in milk IS900 PCR and Nano test, respectively. Of 26 bovine milk samples, 10 (38.5%) were culture positive. Of these 10 milk samples, 8 (30.8%) and 9 (34.6%), were positive in Milk PCR and Nano test, respectively. Sensitivity and specificity of the nano test with respect to milk culture was 91.7 and 96.0%, respectively. Whereas, it was 90.0% (sensitivity) and 92.6% (specificity), with respect to IS900 PCR. In bovine milk samples, sensitivity and specificity of nano test with respect to milk culture was 90.0 and 93.7%, respectively. But sensitivity and specificity was 88.9 and 94.1% with respect to IS900 PCR.

Simplicity and efficiency of this novel 'nano test' makes it suitable for wide scale screening of milk samples in the field. Standardization, validation and re-usability of test has been successfully achieved in field samples. Test was highly specific, simple to perform and easy to be read by naked eyes and does not require laboratory support It can be used as mass screening test involving large number of samples.

4.2 ESTIMATION OF PERFORMANCE CHARACTERISTICS OF ANALYTICAL METHODS FOR MYCOBACTERIUM AVIUM PARATUBERCULOSIS (MAP) DETECTION IN MILK.

Butot Sophie[1], Ricchi Matteo[3], Sevilla Iker[2], Michot Lise[1], Molina Elena[2], Tello Maitane[2], Russo Simone[3], Tomas David[1].

[1]*Nesté Research Center*
[2]*NEIKER*
[3]*IZSLER*

Several methods and technologies are published and applied to detect MAP but there is a lack of validation data and performance characteristics obtained following an international recognized approach (ISO 16140-2:2017) to demonstrate these methods are fit-for- purpose.

The objective was to estimate performance characteristics for three analytical methods used for MAP detection in three types of milk (heat-treated, powdered and raw milk).

Two de-clumping protocols (filtration or sonication) were tested. For each type of milk, samples were spiked with a reference MAP strain at two levels. Samples were stored at -20 °C and distributed to two ISO 17025 accredited laboratories for analysis using three different methods.

Both de-clumping protocols showed good performance by microscopy observation. Close results were obtained by culture and phage assay without PMS with value 0.5 Log cells/ml higher for filtration. Before spiking, MAP levels of four pure cultures were estimated by microscopy (Neubauer chamber), phage assay without PMS and Real Time PCR (f57 and IS900) and used to define the spiking levels. MAP levels in milk samples after spiking were estimated by Most Probably Number (MPN) using Real Time PCR (IS900). Values for low levels ranged from 2,500 to 3,100, 7,500 to 15,000 and 7,500 to 48,000 MPN/50ml for heat treated milk, powdered milk and raw milk respectively. Sensitivity was estimated by comparing reference values with the results obtained by laboratories performing the three analytical methods: culture, Real Time PCR (IS900 and f57) and Peptide magnetic separation (PMS) + phage assay.

Spiked samples showed good homogeneity and stability. Preliminary results obtained in two accredited laboratories showed different performance characteristics (sensitivity) for the three assays tested for MAP detection.

4.3 EFFECT OF CALCIUM HYDROGEN CARBONATE MESOSCOPIC CRYSTALS TO MYCOBACTERIUM AVIUM SUBSP. PARATUBERCULOSIS.

Eiichi Momotani, Rumiko Onishi, Koichi Furusaki, Takashi Onodera

To control and eradicate MAP infection, diagnosis, management for animals, disinfection of MAP in environment including foods are important. In the present study, the bactericidal effect of a novel electrically charged disinfectant, CAC-717 was evaluated. CAC-717 is produced by applying an electric field to mineral water containing calcium hydrogen carbonate to generate mesoscopic crystals. The pH value is 12.39. But once dropped on human skin, the strong alkaline nature was dramatically altered to neutral pH, 8.84 and safe. MAP is under clouds of suspicion in relation to human auto-immune diseases, such as human Crohn's disease and multiple sclerosis. The solution of CAC-717 seems to inactivate MAP by saponification with strong alkalinity.

To control and eradicate MAP infection.

Linda strain of MAP (ATCC43015) was purchased from ATCC and cultured in 7H9 broth with OADC enrichment, glycerin and 0.025% Tween80 for over 3 weeks under mild shaking culture. The bacterial pellet was washed with PBS and distillated water with
0.025 Tween 80. Then culture suspension was mixed with different concentration of CAC-717 and time. Effect of the presence of protein was tested by using albumin. After the treatment, bacterial suspension was mixed with Calsein AM solution and measured the viability by CytoFluor 2300 Fluorescence Measurement System. CFU of the treated MAP was also counted. Cytological observation was carried out light microscopy with Ziehl-Neelsen stain and immunocytology, and by electron-microscope. The coexisting protein was evaluated.

A novel electrically charged disinfectant, CAC-717 by degenerate the cell wall by the alkalinity. The action was not decreased coexisting protein.

This solution may be useful as disinfectant for MAP on certain environmental conditions.

4.4 FREE-LIVING AMOEBAE AS AN ENVIRONMENTAL HOST FOR MYCOBACTERIUM AVIUM SUBSP. PARATUBERCULOSIS.

Hechard Yann[1], **Ascel Samba-Louaka**[1], Etiene Robino[1], Thierry Cochard[2], Willy Aucher[1], Franck Biet[2].

[1]*University of Poitiers*
[2]*INRA Nouzilly*

Free-living amoebae are protozoa found in water and soil where they feed on bacteria. Bacteria are usually digested by phagocytosis however it has been demonstrated that some bacteria (such as Legionella pneumophila or some Mycobacteria) could resist phagocytosis. Free-living amoebae are described as reservoirs of pathogenic bacteria in the environment since bacteria able to survive within these amoebae would also survive phagocytosis from immune cells.

Our objective was to assess to role of free-living amoebae as environmental reservoirs of *Mycobacterium avium subsp. paratuberculosis* (MAP).

The study was performed 1) by studying growth of MAPs within a model amoeba and 2) by isolating environmental amoebae from farms positives for paratuberculosis to look for their infection by MAP.

The results indicate that various MAP strains were able to grow within Acanthamoeba and that they can survive for several days within their host in a phagosome. Importantly, a MAP strain was detected within an environmental amoeba, identified as related to the poorly described Rosculus genus. The bacterial strain was genotyped, showing that it was similar to previous infectious strains isolated from cattle. Rosculus has been described as coprophilic thus we could hypothesize that this amoeba could be found in cattle faeces.

In conclusion, we described that various MAP strains could grow or persist within amoebae and, for the first time, that these bacteria could be found on farm within amoebae isolated from the cattle environment. It validates that amoebae might be a reservoir and vector for the transmission of MAP and stimulates for further studies.

4.5 MYCOBACTERIUM AVIUM SUBSPECIES PARATUBERCULOSIS AND MYELIN BASIC PROTEIN SPECIFIC EPITOPES ARE HIGHLY RECOGNIZED BY SERA FROM PATIENTS WITH NEUROMYELITIS OPTICA SPECTRUM DISORDER.

Marco Bo[1], Giannina Arru[2], Magdalena Niegowska[1], Gianpietro Sehi[2], Leonardo Sechi[1].

[1]Department of Biomedical Sciences, University of Sassari, Sassari, Viale San Pietro 43b, Italy
[2]Department of Clinical and Experimental Medicine, University of Sassari, Italy.

Neuromyelitis optica spectrum disorder (NMOSD) is an immune-mediated inflammatory disease caused by unknown environmental factors acting in genetically susceptible persons. Epstein-Barr virus (EBV) is a postulated contributing agent given its supposed association with multiple sclerosis (MS) and similarity of neurological symptoms between the two disorders. In previous studies we detected a significantly increased prevalence of Abs against peptides derived from *Mycobacterium avium subsp. paratuberculosis* (MAP) homologous to EBV and human epitopes in MS patients.

In this study, we aimeded at investigating whether seroreactivity to these antigens display specific pattern in patients with NMOSD.

Sera of 34 NMOSD Italian patients within one year from diagnosis and 38 age-/sex-matched healthy controls (HCs) were tested for the presence of Abs against peptides derived from EBV (BOLF1305-320, EBNA1400–413), MAP (MAP_402718-32, MAP_0106c121–132) and their human homologs (MBP85–98, IRF5424-434) through ELISA.

NMOSD patients showed elevated levels of antibodies against MAP and MBP compared to healthy controls (44% vs. 5%, p<0.0002 and 50% vs. 2%, p<0.0001, respectively), whereas no difference was observed against EBV, unlike in MS.

Our data highlight for the first time a significant association of MAP with NMOSD. Since the only available test (anti-aquaporin 4) used to diagnose NMOSD gives positive results in at most 80% of patients, the assessed epitopes may be suitable as additional biomarkers for better differentiation between NMOSD and MS. Further studies in a larger cohort are required to confirm these results.

4. 74 BIO-LOAD AND BIO-TYPE PROFILE OF MYCOBACTERIUM AVIUM SUBSP. PARATUBERCULOSIS IN RAW MILK AND COMMERCIAL MILK PRODUCTS USING 6 ANTIGEN AND ANTIBODY BASED TESTS

Manju Singh[1], Shoor Vir Singh[1], Sachin Digamber Audarya[2], Shaswati Pany[3], Adya Prakash Rath[3], Sangram Biswal[3].

[1]*Animal Health Division, Central Institute for Research on Goats, Mathura, Uttar Pradesh, India.*
[2]*Department of Microbiology, College of Veterinary Sciences, Mhow, Madhya Pradesh, India.*
[3]*Department of Epidemiology and Preventive Medicine, College of Veterinary Science & Animal Husbandry, Odisha University of Agriculture & technology, Bhubaneswar*

In India dairy products received no attention as vehicles for transmission of MAP bacilli to human population. Since MAP is not in- activated during pasteurization, there is increased compromise on public health concerns on MAP as 'food borne pathogen'.

To estimate the bio-load and bio-type profile of M. paratuberculosis in raw milk and commercial milk products.

Raw milk of 465 lactating goats from farm and farmers herds, 138 milk products and 133 commercial pasteurized milk (liquid milk-100, flavoured milk-19, milk powder-14) of leading commercial brands sold in the local markets were screened to estimate bio- load of MAP using 6 tests.

Of 465 goat raw milk screened, bio-load of MAP was 13.9-48.8% using antigen (48.8% -i_FAT, 46.6% - microscopy, 13.9% - IS900 PCR) and 39.1-57.4% (39.1- i_ELISA, 57.4%- d_ELISA, 55.6%- LAT) in antibody detection tests. In bio-typing using IS1311 PCR_REA, 72.3% were 'Indian Bison Type'. Except IS900 PCR, 5 tests were comparable and showed high (39.1-57.4%) bio-load of MAP. 'Indian Bison Type' was major bio-type of MAP in lactating goats. Similar findings were reported from other livestock species, in this part of the country. Of 133 commercial milk screened, 42.8, 58.6, 9.0, 27.0, 49.6 and 42.8% were positive in microscopy, i_FAT, IS900 PCR, i_ELISA, d_ELISA and LAT, respectively. i_FAT was most sensitive test as compared to microscopy, i_ELISA and IS900 PCR for estimating bio-incidence of MAP in pasteurized milk. Of 138 commercial milk products screened, 41.3, 39.8, 7.9, 22.4, 33.3 and 23.1% were positive in microscopy, i_FAT, IS900 PCR, i_ELISA, d_ELISA and LAT, respectively.

Microscopy was most sensitive and significantly superior to other tests. Milk was convenient sample in collection and screening. First time in our studies whole milk was used directly as 'test sample'. Presence of live bacilli in raw milk and milk products, pose major threat to human health. High bio-load of MAP in commercial milk products indicate need for implementation of JD control programs at the National level and lower threat to human health.

4.75 DETECTION AND ISOLATION OF MAP FROM SKELETAL MUSCLE TISSUE OF 39 % OF 143 CATTLE AT AN ABATTOIR

HeinrichDahmen[1], Dr. Marc Duenner[1], Dr. Karl Zimmer[2], Dr. Dirk Steinhauer[2], Prof. John Hermon-Taylor[3], Neil Rayment[3].

[1]*Tieraerztliche Praxis Dres. Dahmen & Duenner, Pruem, Germany*
[2]*Landesuntersuchungsamt Rhld.-Pfz., Abt. Tiermedizin, Koblenz, Germany*
[3]*Div. of Diabetes and Nutritional Sciences, King´s College London, London SE1 9 NH, London (UK)*

Paratuberculosis or Johne´s Disease is a worlwide systemic and intestinal infection of ruminants baused by MAP (*Mycobacterium avium* subsp. *paratuberculosis*). The disease is characterized in cattle by a long, sometimes years, incubation time and chronic progression resulting in leaning and quenchless diarrhea. The MAP associated host spectrum covers principally ruminants, including cattle especially dairy herds, sheep, ghoats, camelids, and other species but primates are also susceptable. A link in between MAP and human Crohn´s disease has been discussed for more than 100 years. As seen in previous studies milk and meat could be a scource of MAP infection for humans.

The detection of MAP infected cattle during official meat inspection at an abattoir and isolation of MAP in bovine skeletal muscle by culture and immunofluorescent tests

From 2015 to 2017, fresh samples of faeces, gut tissue and mesenteric lymph nodes as well as different skeletal muscles were collected during officicial meet inspection from 143 cattle with early to advanced pathological signs of MAP infection at a single German abattoir. Samples were taken immediately to the official state´s laboratory in Koblenz, Germany. Faeces, gut wall and associated mesenteric lymph nodes were tested for MAP using ZN staining and histopathology. Skeletal muscle samples from all 143 animals were also tested separately for MAP by culture and PCR and were positive for MAP in 39 %. An initial immunofluorescent study using 2 mouse monoclonal antibodies specific for MAP was carried out blind on 5 muscle culture positive and 5 muscle culture negative muscle samples and was positive in 9 of the 10 cattle.

1. Official meat inspection can detect MAP infected cattle 2. MAP has been detected by culture in different skeletal muscles in 39 % of MAP infected cattle 3. An initial immunofluorescent study using 2 mouse monoclonal antibodies specific for MAP on ten muscle samples was positive in 9 of 10 cattle

The results of the study so far are consistent with the known systemic distribution of MAP but show a higher than expected involvement of skeletal muscle with MAP pathogens.

4.76 ANAEROBIC FERMENTATION OF SLURRY IN MESOPHILIC BIOGAS PLANTS REDUCES *MYCOBACTERIUM AVIUM SUBSP. PARATUBERCULOSIS* IN THE DIGESTATE

Donat Karsten[2], Susannne W. Eisenberg[1], Esra Einax[2], Gerd Reinhold[3], Veit Zoche-Golob[2].

[1]*Animal Diseases Fund of Lower Saxony*
[2] *Thuringian Animal Diseases Fund, Animal Health Service*
[3] *Thuringian State Institute for Agriculture*

Fecal shedding of *Mycobacterium avium subsp. paratuberculosis* (MAP) leads to considerable concentrations of MAP in slurry. The environmental contamination with MAP from positive herds has been postulated to play a role in MAP transmission when this slurry is used as fertilizer. For other infectious agents, anaerobic digestion has shown to reduce bacterial load.

This study aimed at investigation the impact of anaerobic digestion in mesophilic biogas plants on MAP survival in digestate.

A field study included sixteen dairy farms with a known MAP history (estimated apparent within-herd prevalence of MAP shedders at sampling 0.7-21.2%) operating an on-farm biogas plant that processed manure originating from cattle at different proportions. Fecal culture and a commercial quantitative MAP IS900 qPCR were applied on samples collected before, during and after fermentation (mean fermentation temperature 39-45 °C).

Unfermented slurry samples of most farms were MAP culture and qPCR positive. After fermentation, in most samples MAP could no longer be cultured with the exception of two samples from one farm with a high number of shedders at time of sampling. Using a Bayesian binomial model, the probability of getting a MAP negative culture result was predicted to be 93%. In contrast, MAP DNA was detected in most samples when using the IS900 qPCR, resulting in an estimate of 27% for the probability of a negative result in qPCR.

In conclusion, exposing MAP positive slurry to anaerobic digestion in on-farm biogas plants may contribute to a considerable reduction of viable MAP in the digestate. Whether the MAP DNA detected in the fermentation products belongs to degradation products of MAP or to viable MAP, which are either below the detection limit or in a dormant state, cannot be differentiated. Using digestion products as fertilizer instead of slurry probably reduces the risk of MAP transmission to other farms and the risk of re- introduction of the infectious agent by feed.

4.77 MAP IN DAIRY GOAT COLOSTRUM AND MILK

Karianne Lievaart-Peterson[1], Saskia Luttikholt[1], Maarten F. Weber[1], Maaike Gonggrijp[1], Robin Ruuls[2], Ad Koets[3].

[1]*GD Animal Health, Deventer, The Netherlands*
[2]*Department of Bacteriology and Epidemiology, Wageningen Bioveterinary Research, Lelystad, The Netherlands*
[3]*Department of Bacteriology and Epidemiology, Wageningen Bioveterinary Research, Lelystad & Department of Farm Animal Health, Faculty of Veterinary Medic*

Mycobacterium avium subsp. *paratuberculosis (*Map) infection is estimated to be present in 78% of Dutch dairy goat herds. To reduce transmission, it is advocated to snatch kids at birth and feed them cow- and/or artificial colostrum.

This research evaluated the potential risk of colostrum and milk feeds to goat kids within Map test-positive dairy goat herds.

Convenient colostrum (n=117, six herds) and milk samples (n=202), as well as colostrum (n=22) from goats with clinical paratuberculosis were collected. Furthermore twenty nine clinically Map infected cull goats (three herds) were post-mortemed and bilateral milk as well as tissue sampled. Tissues pools consisted of ilium, jejunum, ileocecal valve, mesenteric lymph nodes as well as bilateral mammary tissue and mammary lymph nodes. All milk and colostrum samples were tested by ELISA (ID Vet) and PCR (IS900). Tissue was tested by PCR.

Of the milk samples 47% were antibody positive, but only 4% of these samples were PCR positive. In the initial 117 colostrum samples there was no Map signal (0%). Two of the colostrum samples from the goats with clinical paratuberculosis were PCR- positive. There was no one-on-one match between antibody and DNA positives. Results from sacrificed goats showed a PCR signal in 72% in the intestine pools and 83% in the mesenteric lymph nodes. In 34% of the mammary lymph nodes a PCR signal was seen as well as 45% in the mammary tissue. Of those mammary tissue PCR positives, only 15% were PCR positive in their milk plus three doubtful PCR (Ct>36) signals.

The antibody responses in a high proportion of milk and colostrum samples suggested that the selected goats had been in contact with Map and/or been vaccinated. Despite vaccination most of the sacrificed goats had one or more PCR positive tissue sample. The possible risk for Map transmission of Map via milk and colostrum seems to be low.

4.78 *MYCOBACTERIUM AVIUM* SUPESPECIES *PARATUBERCULOSIS* OCCURENCE IN CROHN'S DISEASE PATIENTS AT A BRAZILIAN REFERRAL CENTER FROM 2007 TO 2017

Isis de Freitas Espeschit[2], Antônio Augusto Fonseca Jr.[1], Jéssica Lobo Albuquerque[2], Maria de Lourdes Abreu Ferrari[3], Maria Aparecida Scatamburlo Moreira[2], Isabel Azevedo Carvalho[4].

[1]*Laboratório Nacional Agropecuário, Ministério da Agricultura Pecuária e Abastecimento*
[2]*Universidade Federal de Viçosa*
[3]*Instituto Alpha de Gastroenterologia- Hospital das Clínicas, Universidade Federal de Minas Gerais*
[4]*Universidade Estadual do Maranhão*

Mycobacterium avium subspecies *paratuberculosis* (MAP) is frequently isolated from samples of patients with Crohn's disease (CD), a disease that shares with paratuberculosis, that affects ruminants, diverse clinical and histopathological characteristics. Despite this, the role of MAP in CD pathology has not yet been clarified.

The present study was conducted at the Alpha Institute of Gastroenterology of the Hospital das Clínicas of the Federal University of Minas Gerais, a referral center for the treatment of inflammatory bowel diseases, and aimed to identify the occurrence of MAP over 10 years among patients with CD.

A probabistic sampling (CI = 99%; p = 0.05) was performed, where, from the 421 CD patients that attended the colonoscopy sector between 2007 and 2017, 258 patients were selected, whose intestinal biopsy specimens were recovered from blocks of paraffin filed by the institute and submitted real-time PCR in triplicate, with primers 18S (control), IS900 and F57.

All samples were positive at 18S and of these, 44 (17.05%) were positive for both primers, confirming the presence of MAP DNA.

This is an unprecedented result, coming from an interval of 10 consecutive years and representing the actual occurrence of the bacterium in this population, which instigates the development of research in order to clarify the role of MAP in the appearance and evolution of the disease, a controversial fact so far. Also, in a long term, it can contribute to the elaboration of therapeutic strategies. This study is part of the innovative global strategy for interaction between animal, human and environmental health, One Health.

4.79 DETECTION OF *MYCOBACTERIUM AVIUM* SUBSP. *PARATUBERCULOSIS* IN COMMERCIAL MILK AND MILK PRODUCTS IN ARGENTINA

Karina Cirone[1], Claudia Morsella[1], Laura Mendez[1], Fernando Paolicchi[1].

[1]*Production Animal, National Agricultural Technology Institute, Buenos Aires, Argentina.*

Paratuberculosis (PTBC) or Johne´s disease is Chronic Enteritis disease caused by *Mycobacterium avium* subsp. *paratuberculosis* (MAP) which affects cattle and other animal species. MAP has been linked to Chronic Inflammatory Bowel disease in humans, named Crohn's disease. Milk and milk products are possible sources of infection to humans as it was suggested that this mycobacterium could resist pasteurization conditions.

The aim of this work was to detect MAP and other mycobacteria in commercial milk and processed dairy products by bacteriological culture and polymerase chain reaction (PCR).

Three hundred and eighty four samples of commercialized milk, 24 samples of yogurts and 24 samples of cheeses in dairy Cuenca Mar y Sierras, Argentina, were tested along 24 months using culture in specific media (Herrold plus mycobactina J, Stonebrink and Loweinsten Jensen) without decontamination and conventional IS900 PCR.

All samples were negative for growing MAP and other mycobacteria using the techniques described. However, from the commercial milk, 1.56% (6/384) of the samples was positive for IS900 PCR. The PCR for yogurts and cheeses samples was 100% negative.

The positives samples by IS900 PCR indicate that MAP is present in raw milk and could be killed during the pasteurization. The PTBC is a disease that is present in the Argentinean dairy herds and, in consequence, MAP could be found in raw milk. Control policies should emphasize more in hygiene, cold chain, etc. not only inside the industries, but also in points of sale to make sure that consumers will purchase a product with a high quality.

4.80 MAP AND ZNT8 PEPTIDES INDUCE IMMUNE RESPONSES IN PBMC OF T1D PATIENTS

Magdalena Niegowska[4], Izabela Kubiszewska[1], Lidia Gackowska[1], Ewa Balas[2], Joanna Trojanek[2], Mieczyslaw Szalecki[3], Jacek Michalkiewicz[2], **Leonardo A. Sechi[4]**

[1]*Department of Immunology, Collegium Medicum in Bydgoszcz, Nicolaus Copernicus University in Torun, Bydgoszcz, Poland*
[2]*Department of Clinical Microbiology and Immunology, The Children's Memorial Health Institute, Warsaw, Poland*
[3]*Clinic of Endocrinology and Diabetology, The Children's Memorial Health Institute, Warsaw, Poland*
[4]*Department of Biomedical Sciences, Division of Microbiology and Virology, University of Sassari, Italy*

The role of *Mycobacterium avium* subsp. *paratuberculosis* (MAP) as a putative environmental agent triggering or accelerating type 1 diabetes (T1D) has been hypothesized based on an elevated responsiveness to MAP-derived epitopes and homologous human antigens in Italian T1D populations. The present study was performed to evaluate differences in the recognition of the same peptide set and immune cell responses in a population of T1D patients with a distinct biogeographical background.

The present study was performed to evaluate differences in the recognition of the same peptide set and immune cell responses in a population of T1D patients with a distinct biogeographical background.

Sera of 75 Polish T1D pediatric subjects and 26 age-matched controls (HC) were tested for the presence of antibodies against MAP peptides and homologous human epitopes corresponding to proinsulin (PI) and ZnT8 fragments. PBMCs of a pilot cohort (n=17 T1D, n=4 HC) underwent stimulation with MAP3865c133–141 or ZnT8186-194 peptide followed by the evaluation of cytokine levels (IL-1β, IL12p40, TNFα, INFγ) through ELISA and immune cell populations using flow cytometry.

Compared to HC, seroreactivity among T1D showed increased levels for MAP/PI epitopes (respectively: 27% vs. 11.54%, $p<0.008$ and 28.38% vs. 7.69%, $p<0.001$) and, additionally, ZnT8 in males (29.63% vs. 10%, $p<0.043$). In PBMC culture, both MAP and ZnT8 peptides induced expression of IL-1β and IL12p40 in T1D subjects, while TNFα was observed at decreased levels. For IL-1β and INFγ, high concentrations corresponded to T1D subjects bearing anti-MAP antibodies who also presented increased cell apoptosis, CD3 and CD14+/CD16+ monocyte levels. Early-life diet based on breastfeeding and milk derivatives moderately correlated with the antibody status.

Seroreactivity to MAP/human antigens is increased in T1D subjects of populations with distinct environmental and cultural background. Good correlation between values obtained for the homologous peptide pairs points at cross-reactivity through which mechanisms of self tolerance may be disrupted leading to autoimmunity. Cell responses to stimulation with MAP-derived peptides should be evaluated in a larger cohort.levels. Early-life diet based on breastfeeding and milk derivatives moderately correlated with the antibody status.

Richard S. Merkal Award Abstracts

DETECTION OF MARKER-SPECFIC IMMUNE RESPONSESIN CALVES AGAINST A NOVEL MARKED MAP PARENT VACCINE STRAIN

[1]**Lucy Luo,** Joroen De Buck

[1]*University of Calgary.*

The control of Mycobacterium avium subsp. paratuberculosis (MAP) infection has been hindered by poor diagnostic sensitivity and lack of suitable vaccines. Current vaccines are ineffective in preventing infection and lack the ability to differentiate vaccinated from naturally infected animals with Johne's disease (JD) or bovine TB.

We aimed to develop a novel marked MAP vaccine strain capable of eliciting marker-specific immune responses in vivo, allowing the differentiation between non-infected, infected and vaccinated animals (DIVA).

We created a positive marker strain by incorporating a foreign antigen in a scaffold MAP protein (PepA) using allelic exchange. To find the best epitope candidate for this marker, we inserted 7 unique epitopes in 3 different locations in PepA and assessed mycobacterial expression. Cellular immune responses against the markers were assessed in vivo using a calf infection model, where calves were inoculated with either the marked strain (n=6), a WT field strain (n=6), or uninfected controls (n=4). Whole blood was collected at 3w intervals until necropsy at 4.5 months PI. Marker-specific IFN-g release was detected by stimulating whole blood with the selected synthetic peptides or their scrambled (-S) versions, followed by an IFN-gELISA.

Out of 21 unique PepA-epitope combinations, the HA epitope (a 8aa hemagglutinin peptide) was selected based on high expression in M. smegmatis. This PepA-HA protein was secreted by the genetically modified MAP strain. To demonstrate successful infection, all WT- and HA-infected calves were confirmed tissue culture (+), while uninfected controls remained culture (-). HA-S-stimulated samples showed a significant IFN-g response in HA-infected calves compared to WT-infected (p=0.016) and uninfected (p=0.019) groups at 4.5 months PI.

This study suggests that HA as a positive vaccine marker, coupled with HA-S stimulation in an IFN-g release assay, holds potential as a valuable diagnostic tool in creating a DIVA JD vaccine.

PREVALENCE OF MYCOBACTERIUM AVIUM SUBSP. PARATUBERCULOSIS ACROSS CANADA, BASED ON ENVIRONMENTAL SAMPLING

Caroline S. Corbett[1], S. Ali Naqvi[1], Cathy Bauman[2], Jeroen De Buck[1], Karin Orsel[1], David F. Kelton[2], Herman W. Barkema[1].

[1]*University of Calgary.*
[2]*University of Guelph*

Accurately estimating the prevalence of Mycobacterium avium subsp. paratuberculosis (MAP) infections is an important part of controlling the spread of infections and monitoring the effectiveness of control programs. Due to the variety of diagnostic methods that are used across Canada, prevalence estimates between regions and programs cannot be compared.

The aim of the current study was to estimate the prevalence of MAP infection in Western Canada, Ontario, Quebec, and the Atlantic provinces using the same sampling method across Canada.

In all 10 provinces, 2 environmental samples were collected from adult cattle areas on 362 dairy farms and cultured for the detection of MAP. Diagnostic sensitivity (Se) and specificity (Sp) were calculated for the ability to detect positive farms using only 2 environmental samples, which were applied to the results in order to accurately estimate the true prevalence in the 4 regions of Canada. The Se and Sp calculated to be 0.48 and 0.99, respectively.

These test characteristics were applied to the environmental culture results from the 362 participating farms in the 4 regions resulting in adjusted true prevalence estimates of 67% of farms in Western Canada, 58% in Ontario, 22% in Quebec, and 41% in the Atlantic Provinces.

This is the first nation-wide study in which the same sampling method was used to estimate prevalence of MAP-infected farms across all of Canada, allowing for a direct comparison among regions. Future research will allow for the investigation of factors that contribute to the differences in prevalence observed in the current study, and lead to better surveillance and impact of current control programs across the country.

Foro Iberoamericano de Paratuberculosis

Comité organizador

Gilberto Chávez Gris (Chairman)
Presidente de la Sociedad Mexicana de Paratuberculosis y otras micobacteriosis.
Medicina Veterinaria y Zootecnia.
Universidad Nacional Autónoma de México
gris@unam.mx

Edith Maldonado Castro
Tesorera de la Sociedad Mexicana de Paratuberculosis y otras micobacteriosis.
Medicina Veterinaria y Zootecnia.
Universidad Nacional Autónoma de México
macae09@hotmail.com

Leticia Carolina García Sánchez
Socia/Fundadora de la Sociedad Mexicana de Paratuberculosis y otras micobacteriosis.
Medicina Veterinaria y Zootecnia.
Universidad Nacional Autónoma de México
caro021292@hotmail.com

Ana Laura Hernández Reyes
Socia/Fundadora de la Sociedad Mexicana de Paratuberculosis y otras micobacteriosis.
Medicina Veterinaria y Zootecnia.
Universidad Nacional Autónoma de México
anahdzreyes@gmail.com

Programa Foro Iberoamericano
(por invitación)

14th INTERNATIONAL COLLOQUIUM ON PARAUBERCULOSIS, RIVIERA MAYA, QUINTANA ROO, MÉXICO. 4 y 7 de junio de 2018

Lunes 4 de junio	
Horario: 12:50-14:50	
12:50-13:00	Bienvenida
13:00-13:20	**Argentina** PARATUBERCULOSIS EN LA ARGENTINA. SITUACIÓN ACTUAL, AVANCES E INVESTIGACIONES. Coordinador: Fernando Alberto Paolicchi.
13:20-13:40	**Brasil** AVANCES Y DESAFÍOS DE LA PARATUBERCULOSIS EN BRASIL. Coordinadora: Maria Aparecida Scatamburlo Moreira.
13:40-13:50	RECESO
13:50-14:10	**Chile** PASADO, PRESENTE Y FUTURO DE LA PARATUBERCULOSIS EN CHILE. Coordinador: Miguel Salgado
14:10-14:30	**Colombia** PARATUBERCULOSIS IN COLOMBIA: PAST, PRESENT AND FUTURE. Coordinador: Jorge A. Fernández-Silva
14:30-14:50	COMENTARIOS
Miércoles 7 de junio	
Horario: 8:00-10:00	
8:00-8:20	**España** LA PARATUBERCULOSIS EN ESPAÑA, ABORDAJE DEL CONTROL DE LA ENFERMEDAD Y PRINCIPALES LÍNEAS DE INVESTIGACIÓN. Coordinador: Joseba M. Garrido
8:20-8:40	**México** PARATUBERCULOSIS EN MÉXICO, AVANCES, PERSPECTIVAS Y RETOS Coordinador: Gilberto Chávez Gris
8:40-9:00	COMENTARIOS
9:00-9:15	RECESO
9:15-9:20	CONCLUSIONES DEL FORO IBEROAMERICANO

Lista de Participantes al Foro Iberoamericano:

I. ARGENTINA (4)

1. **Fernando Paolichi***
2. María Fiorella Alvarado Pinedo
3. Gabriel Eduardo Travería
4. María de la Paz Santangelo

II. BRASIL (1)

5. **María Aparecida Scatamburlo Moreira***

III. COLOMBIA (5)

6. **Jorge Arturo Fernández Silva***
7. Nathalia Correa Valencia
8. Jhon Jairo Tuberquia Londoño
9. José Miguel Hernández Agudelo
10. María de los Ángeles Largo Quintero

IV. CHILE (2)

11. **Miguel Salgado Alfaro***
12. Pamela Steuer Traeger

V. ESPAÑA (3)

13. **Joseba M. Garrido Urkullu***
14. Marta Alonso
15. Natalia Ezguezabal Vega

VI. MÉXICO (12)

16. **Gilberto Chávez Gris***
17. Edith Maldonado Castro
18. Ana Laura Hernández Reyes
19. José Ramón González Sáenz Pardo
20. Luisa Ibarra Lemas
21. Thalia Xóchitl Torres Barajas
22. Rosa Berta Angulo Mejorada
23. Antonio Ortiz Hernández
24. Gilberto Ballesteros Rodea
25. Gabriel Pallás Guzmán
26. Victoria E. Castrellón Ahumada

Moderadores: Gilberto Chavez Gris, Edith Maldonado Castro

Argentina

PARATUBERCULOSIS EN LA ARGENTINA. SITUACIÓN ACTUAL, AVANCES E INVESTIGACIONES. AÑO 2018

Responsable de la información: Fernando Alberto Paolicchi. Jefe de Grupo de Sanidad Animal, Responsable de Laboratorio de Bacteriologia, EEA-INTA Balcarce, Argentina y Profesor Asociado en Facultad de Ciencias Agrarias, Universidad Nacional de Mar del Plata.

Ministerio de Agroindustria
Presidencia de la Nación

-Laboratorio Bacteriología, EEA INTA Balcarce y Facultad de Ciencias Agrarias, Universidad Nacional de Mar del Plata. Equipo de trabajo: FA.Paolicchi (*responsable de la información*), CG.Morsella, L.Mendez, MA.Fiorentino, B.Vasini, F.Bresky, K.Cirone, V.Salazar. *paolicchi.fernando@inta.gob.ar*

En la Argentina, la paratuberculosis fue descripta por primera vez en bovinos por el Dr. F. Rosenbusch en el año 1935 y en años posteriores en otras especies. La identificación de *M. paratuberculosis* fue realizada décadas posteriores y nuestro grupo de investigación de INTA Balcarce ha desarrollado actividades de investigación, diagnóstico y transferencia desde el año 1985. Nuestro laboratorio se encuentra ubicado en la zona denominada Pampa Húmeda, Provincia de Buenos Aires en relación a cuencas lecheras y campos de cría y engorde de bovinos y en menor proporción de establecimientos lecheros y de carne ovina y caprina y cría de ciervos para carne.

-Los objetivos generales del grupo son:

•Diagnostico serológico, bacteriológico, patológico y por inmunidad celular en bovinos, ovinos, caprinos y ciervos con PTB
•Desarrollo de propuestas de control y seguimiento de rodeos infectados con PTB
•Estudios epidemiológicos a nivel nacional para conocer el estatus de la enfermedad en diferentes sistemas de cría y leche en bovinos
•Investigar las características inmunopatogenicas, epidemiológicas y de la respuesta inmune en bovinos, ovinos caprinos y ciervos en confinamiento para el control de la enfermedad
•Mantenimiento de un cepario nacional mediante la colección de micobacterias tipificadas por métodos moleculares.
•Caracterización y secuenciación de cepas de MAP con virulencia diferencial para estudios de patogenia y desarrollo de estrategias diagnósticas.
•Ensayos de antigenicidad y virulencia de cepas de micobacterias ambientales y su interferencia con Paratuberculosis y Tuberculosis en bovinos.
•Ensayos de virulencia de cepas de MAP en modelo bovino para estudios de inmunidad celular, humoral y de dinámica de excreción en bovinos.
•Identificación de la presencia de MAP en productos alimenticios derivados de producción bovina y caprina

•Docencia de grado y de posgrado en programas de capacitación para especializaciones, maestrías y doctorados de alumnos del país y extranjeros.
•Cursos de entrenamiento para profesionales y técnicos de diversos orígenes.

Servicio Diagnostico Veterinario Especializado (SDVE): Nuestro grupo de trabajo en Paratuberculosis compuesto por 4 profesionales y 4 técnicos especializados, realiza diagnóstico y transferencia a veterinarios y al sector agropecuario para entregar el diagnostico de situación epidemiológica en los establecimientos infectados con MAP. Hasta el año 2017 se ha generado información a través de la realización de aproximadamente 270.000 análisis serológicos mediante ELISA *in house*, de las cuales 156.883 muestras de suero analizadas epidemiológicamente pertenecieron a 7 especies animales. Estas muestras provenían de 120 partidos y/o departamentos de 7 provincias argentinas. Tres especies fueron las que contribuyeron con el 99,8 % de las muestras: mayor porcentaje correspondió a bovinos (*Bos Taurus*), en segundo lugar a ciervos (*Cervus elaphus*) y el tercer lugar a caprinos (*Ovis capris*). El resto fueron muestras de llama (*Lama glama*), ovino (*Ovis aries*), muflón (*Ovis musimon*) y búfalo (*Bubalis bubalus*) determinando para nuestro país y principalmente para la región un análisis retrospectivo para las diferentes especies en un periodo de 26 años, con registros de fecha, sistema de producción, nombre del establecimiento y el distrito y provincia de origen. Además se realizó el análisis temporo espacial para cada sistema productivo y cada especie afectada de PTBC utilizando el modelo Bernoulli. El cultivo bacteriológico utilizando diferentes medios de cultivo y la caracterización de la histopatología, son herramientas del diagnóstico de rutina y especializado para la región.

Ensayos de Infección Experimental: desarrollamos modelos de infeccion experimental *in vivo* en bovinos y también en cultivos celulares para estudios de patogenia y virulencia de cepas de micobacterias, especialmente MAP y micobacterias ambientales, con lo cual se desarrollan Tesis Doctorales. Asimismo hemos estudiado los efectos de la vacunación con antígenos micobacterianos en terneros, el desafío de bovinos y ovinos con cepas de diversos orígenes y características genotípicas.

Cepario de Micobacterias: el cepario perteneciente a la Colección de Bacterias de Enfermedades en Veterinaria de relevancia para la institución (Red de Germoplasma de INTA) mantiene más de 345 cepas de MAP todas aisladas a partir de muestras de materia fecal, leche, productos lácteos procesados en Argentina. Más de 100 cepas han sido tipificadas mediante MLVA (MIRUs-VNTR) y obtenido el esquema de relación genotípica por análisis goeBurst, identificando su origen, el sistema productivo asociado a cada uno de los diferentes genotipos. Las cepas se han seleccionado para la secuenciación completa en el contexto colaborativo de un proyecto internacional y para estudios de virulencia en modelos bovinos y en células en cultivo para identificación de interleucinas como respuesta inmune.

Estudios en alimentos infectados con MAP: hemos estudiado en los últimos 8 años la presencia de MAP en leches crudas y leches de consumo, hallando hasta un 2,8% de las leches pasteurizadas y ultra pasteurizadas con presencia de MAP por cultivo y por PCR IS900. Además hemos identificado MAP en carnes de bovinos y de caprinos provenientes de establecimientos con PTBC clínica con el aislamiento de leche y de carnes (cortes de lomo, diafragma e hígado) de animales infectados. Asimismo hemos trabajado con productos lácteos procesados tales como quesos y yogures obteniendo resultados originales frente a la contaminación experimental de los mismos.

Ministerio de Agroindustria
Presidencia de la Nación
INTA

-Instituto de Biotecnología, INTA Castelar. Equipo de trabajo: MP.Santangelo (*responsable de la información*), M.Alonso, ML.Mon, M.Colombatti, MI.Romano, M.Cuerda, D.Moyano, N.Griffa. *santangelo.maria@inta.gob.ar*

El CICVyA INTA Castelar se encuentra ubicado en Morón, Provincia de Buenos Aires a 50 km de la Ciudad de Buenos Aires. Es un centro de investigación en ciencias agronómicas y veterinarias, contando con institutos de investigación y desarrollo de tecnología.

Los objetivos generales del grupo:

- Estudio de bases moleculares de la relación hospedante-patógeno en paratuberculosis bovina
- Desarrollo de nuevas estrategias de diagnóstico y prevención de la enfermedad

Objetivos específicos:

- Diversidad genética del complejo *M.avium* causantes de enfermedad en animales-humanos de Argentina.
- Vacunas contra paratuberculosis
- Relación hospedador-patógeno en infecciones por *Mycobacterium avium* subsp. *paratuberculosis*
- Producción de mutantes atenuadas de *Mycobacterium avium* subsp. *paratuberculosis.*
- Estudio de genes de virulencia.
- Desarrollo de métodos para caracterizar las micobacterias que afectan a cerdos en Argentina
- Identificación de lípidos inmunoreactivos en cepas locales de *M. bovis* y del complejo MAC

Diversidad genética de MAC: El grupo ha descrito la diversidad genética de aislamientos de MAC de diferentes hospedadores con 26 aislamientos de MAH y 61 aislamientos de MAP provenientes de humanos y bovinos, genotipificados por MIRU-VNTR, asignándose un patrón INMV según la base de datos MAC-INMV database *(http://mac-inmv.tours. inra.fr/)*. Esta herramienta fue útil para describir la diversidad genética de MAH y MAP, así como también para identificar 6 nuevos patrones reportados en la base de datos. Otras actividades son el desarrollo de nuevas técnicas para el diagnóstico de Paratuberculosis con el fin de mejorar su control en rodeos para carne como para leche. Se ha avanzado en el desarrollo de una inmunoaglutinación directa con partículas de látex y ELISA utilizando antígenos específicos para la detección de PTBC.

Vacunas contra paratuberculosis. El objetivo de esta línea es evaluar cepas locales para su uso como vacunas. A partir de ensayos realizados por nuestro grupo en modelos murinos, seleccionamos una cepa local de MAP como cepa candidata a vacuna para determinar la respuesta inmune en bovinos. Se realizó un ensayo de vacunación, en bovinos libres de TB y PTB en el campo experimental de INTA. Se evaluó la respuesta inmune frente a la vacuna por medición de anticuerpos específicos en el suero (IgG total e isotipos IgG1 e IgG2), IFN gamma, citoquinas, poblaciones linfocitarias de sangre periférica (por citometria de flujo) e intradermorreacción doble comparativa en tabla del cuello

(PPDa – PPDb). En etapa experimental en ratones estamos evaluando un refuerzo a la vacunación con MVA85. Con el objetivo de comprender los mecanismos de la respuesta inmune frente a la infección con MAP y de identificar marcadores de protección y/o enfermedad, se estudia la evolución de la PTBC en terneros nacidos en establecimientos con PTB, detectando anticuerpos específicos de forma temprana y liberación de IFNg. Se han estudiado mutantes atenuadas de MAP para la identificación de genes de virulencia con el fin de obtener candidatos vacunales, obteniendo mutantes por recombinación homóloga, como la mutante en el gen lprG y avanzando en la obtención de cepas mutantes en *M. tuberculosis* y *M. bovis*, para evaluar su rol en la enfermedad. En un proyecto conjunto de apoyo al desarrollo de las Biotecnologías en el MERCOSUR- BIOTECH II, se trabajó en el desarrollo y validación de pruebas diagnósticas e inmunógenos para el control de las micobacterias que afectan humanos y cerdos. Con muestras obtenidas de producciones de los diferentes países del MERCOSUR, se ensayaron un kit de Myc4TB-VK Real time PCR, diseñado por los investigadores del grupo español para detectar las diferentes especies del género Mycobacterium. Otras actividades son la identificación de lípidos inmunoreactivos en cepas del complejo MAC, para ser utilizados en el diagnóstico de cerdos y bovinos infectados con micobacterias. Con el fin de evaluar la antigenicidad de las fracciones lipídicas, se evaluarán en un ELISA para la selección de la fracción que sea reconocida por el mayor número de sueros positivos.

-Área Microbiología, CEDIVE. Facultad Ciencias Veterinarias. Universidad Nacional de La Plata. Equipo de trabajo: F.Alvarado (*responsable de la información*), G.Travería. MA.Romero, L.Peralta, LA.Di Paolo, P.Sosa, M.Jaureguiberry. *falvarado@fcv.unlp.edu.ar*

El CEDIVE se encuentra ubicado en la región denominada Zona Deprimida del Salado en la provincia de Buenos Aires. La casuística proviene principalmente de bovinos pertenecientes a rodeos de carne o lecheros. Nuestro grupo de docentes investigadores (7) en PTBC desarrollan docencia, investigación, extensión y transferencia

Objetivos generales del grupo:

- Producir reactivos locales de diagnóstico de PTB
- Aplicar propuestas de control en establecimientos con PTB endémica
- Difundir las características patogénicas, epidemiológicas y del control de la PTB

Objetivos específicos:

- Producir antígeno protoplasmático de *Mycobacterium avium* subsp. *avium* para su uso en pruebas serológicas de ELISA y de IDA.
- Producir micobactina para su uso en los medios de cultivo de MAP.
- Elaborar medios de cultivo líquido para acortar los tiempos para la incubación de MAP.
- Determinar el impacto de la presencia de otras micobacterias ambientales en los resultados de las pruebas diagnósticas de PTB.

El grupo realiza trabajos en establecimientos con PTB combinando el uso de pruebas de inmunidad celular e inmunidad humoral y el cultivo de MAP. Están en curso 3 tesis doctorales que tienen como tema principal el diagnóstico y manejo de la PTB. Además se realizan estudios epidemiológicos en rodeos lecheros como parte del plan de trabajo pos doctoral de CONICET. Se trabaja en conjunto con investigadores del INTA Castelar en distintos proyectos resaltando un Proyecto sobre "Evaluación de cepas locales de MAP como candidatos a vacunas contra paratuberculosis", aspirando tener una opción aplicable para la prevención de la PTB. Nuestro grupo realiza charlas informativas de actualización sobre PTB organizadas por Fundaciones y Asociaciones Rurales de nuestra zona, destacando una actividad con la Asociación Rural zonal desde el año 2017 con la cual se solicita a los productores que envían bovinos a remates feria, la presentación de informes serológicos negativos a PTB, demostrando asi la preocupación por el aumento de casos clínicos y muerte de animales debidos a esta enfermedad en la zona de influencia.

-Cátedra de Inmunología, Facultad de Ciencias Veterinarias, Universidad de Buenos Aires. Equipo de trabajo: S.Mundo (*responsable de la información*), B.Fernandez, S.Colavecchia, A.Jolly, G.Ingratta, A.Jar, A.Stempler. *smundo@fvet.uba.ar*

La Facultad de Veterinaria de la Universidad de Buenos Aires se encuentra emplazada en la zona denominada CABA, dentro del ámbito de la Ciudad de Buenos Aires, Provincia de Buenos Aires. Desarrolla actividad principal de docencia de grado y de posgrado y actualmente realiza diagnósticos de PTBC como servicio y principalmente investigación con líneas fundadas en el estudio de la inmunopatología y respuesta inmune para el diagnóstico de la enfermedad.

Objetivos Generales del grupo:

- Docencia de Grado y Posgrado
- Desarrollo de tecnología con base inmunodiagnóstica para PTBC
- Estudios de las características inmunopatogénicas y la relación huésped patógeno.

Los estudios se centran en las nuevas propuestas de control de la enfermedad utilizando herramientas desarrolladas para el diagnostico en suero, leche y materia fecal de bovinos. También la evaluación de la respuesta inmune y sus aportes para el diagnóstico de la enfermedad, son líneas de investigación parte del desarrollo de Tesis Doctorales en el grupo. Con esto se propone estimar la seroprevalencia de la PTBC bovina en ciertas regiones de la Argentina, mejorar la detección de los animales infectados y ampliar el conocimiento sobre los aislamientos locales de MAP. Desarrollar técnicas de identificación de la bacteria o sus componentes en muestras de materia fecal, suero y leche y de la respuesta inmune inducida por la infección en suero y leche para mejorar la detección de los animales infectados con MAP. Para esto se aplican anticuerpos específicos producidos en el propio laboratorio para concentrar las bacterias mediante el uso de perlas inmunomagnéticas y disminuir las interferencias en la técnica de PCR-IS900. También en el grupo se busca la presencia de biomarcadores de infección temprana (ManLam) para lo que se está desarrollando un ELISA de captura. Se estudia la utilización de antígenos de MAP como PPA, p34cx y ManLAM y la identificación de los isotipos IgG, IgG1, IgG2 e IgA en pruebas de ELISA indirecto.

Se ha desarrollado una plataforma para mejorar el diagnóstico de la PTBC y de otras enfermedades infecciosas en animales en función de la producción y aplicación de anticuerpos monoclonales. A su vez se propone la producción de anticuerpos monoclonales anti-IgA bovina y la caracterización y evaluación de anticuerpos monoclonales anti-Map que reconocen a *Mycobacterium avium* subsp. *paratuberculosis* o alguno de sus componentes (Lipoarabinomanano, Proteínas P34 y HP65, entre otras). Los anticuerpos monoclonales anti-IgA bovina se aplican para mejorar la identificación de anticuerpos anti-Map en leche y materia fecal, por ejemplo, a través de la técnica de IgA-ELISA-PPA. Los anticuerpos monoclonales anti-Map se utilizarán en el diagnóstico etiológico de la PTBC a través de la captura del microorganismo.

SENASA

-Laboratorio de Micobacterias. SENASA. Equipo de trabajo: B.Alonso (*responsable de la información*), I.Rodrigo, M.Sanchez. *balonso@senasa.gob.ar*

El Laboratorio Animal del Servicio Nacional de Sanidad Agroalimentaria (SENASA) se encuentra ubicado en el Gran Buenos Aires y es referente en tuberculosis bovina y paratuberculosis de la Organización Mundial de Sanidad Animal (OIE). El Dr Alonso es experto y referente en la OIE en tuberculosis bovina y responsable de la Coordinación de Bacteriología del Laboratorio Animal del SENASA. Este Laboratorio de Referencia, brinda formación científica y técnica al personal de otros países miembros de la OIE y coordina estudios científicos y técnicos en colaboración con otros laboratorios de Argentina (INTA, Universidades, Ministerio de Agroindutria provincial) y el resto del mundo u organizaciones sobre temas relacionados con el diagnóstico y el control de las enfermedades producidas por micobacterias.

Objetivos Generales del grupo:

•Capacitación a Servicios Oficiales de otros países y de Argentina
•Control de Tuberculinas y de test diagnósticos
•Colaboración y coordinación de aspectos inherentes a las campañas de TBC y trabajos en PTBC
•Colaboración en Cursos de Posgrado a nivel nacional e internacional (CEBASEV).

AVANCES Y DESAFÍOS DE LA PARATUBERCULOSIS EN BRASIL

Maria Aparecida Scatamburlo Moreira[1]*, Isis de Freitas Espeschit Braga[1], David Germano Gonçalves Schwarz[2], Isabel Azevedo Carvalho[3]

[1]Universidade Federal de Viçosa (UFV), MG - Brasil
[2]Universidade Federal do Piauí (UFPI), PI - Brasil
[3]Universidade Estadual do Maranhão (UEMA), MA, Brasil
*contacto: masm@ufv.br

Brasil es el quinto país más grande del mundo, con más de 8.515.767 km_2, localizado entre las latitudes 5°16'19" nortey 33°45'07" sur y entre los meridianos 34°47'34" y 73°59'26" oeste. Su litoral está bañado por el Océano Atlántico, limitando con casi todos los países de América del Sur, excepto Chile y Ecuador. Con una gran extensión territorial, un clima diversificado, lluvias regulares, energía solar abundante y casi el 13% de toda el agua dulce disponible en el planeta, Brasil tiene 388 millones de hectáreas de tierras agrícolas fértiles con alta productividad, de las cuales 90 millones aún no fueron explotadas. Todos estos factores contribuyen para que el agronegocio brasilero sea una actividad próspera, segura y rentable. Este escenario es una evolución positiva que se remonta al período de la colonia de Brasil, que se basaba en monocultivos y cría de animales.

A pesar de la recesión económica que vive Brasil, el sector agropecuario es uno de los pocos que ha sostenido índices de crecimiento positivos. La actividad agropecuaria es responsable por el 21% del Producto Interno Bruto (PIB), el 27% de los empleos y el 43% de las exportaciones. A pesar de los buenos resultados, se cree que la ganadería podría tener índices superiores si hubiera mayor compromiso en los servicios de salud animal e incentivos del gobierno. Creemos que una comunicación efectiva entre las partes involucradas y los investigadores podría ayudar en la identificación de los problemas que afectan a la ganadería en la búsqueda de estrategias para resolverlos conjuntamente.

Algunas enfermedades como la Fiebre Aftosa, Brucelosis y Tuberculosis poseen Programas Nacionales instituidos que posibilitaron el control y el direccionamiento para la erradicación de algunas de ellas. Sin embargo, en relación a la Paratuberculosis, aún no existe un programa de control, ni un aparente interés de los órganos competentes, aunque sea de conocimiento la presencia de esta enfermedad en el rebaño brasileño.

La primera identificación de la enfermedad en el país fue registrada en 1915 por Octávio Dupont en bovinos de la raza Flamenga, importados de Bélgica a Río de Janeiro. Desde entonces, se pensó que la enfermedad era exótica y los casos sólo restringidos a los animales importados. Sin embargo, Dacorso Filho y colaboradores, en 1960, identificaron la enfermedad en bovinos nacidos y criados en Brasil. Los registros subsiguientes de la enfermedad fueron verificados en diferentes estados brasileños demostrando que el agente realmente circula entre los rebaños nacionales.

La distribución de la Paratuberculosis en los estados brasileños todavía se basa en diferentes metodologías y en diferentes tipos de muestras, lo que dificulta la comparación y la interpretación de los resultados, así como la detección real de la prevalencia en el país. En una revisión realizada por Espeschit y colaboradores (2017), los autores verificaron que, entre los países de América Latina,

Brasil es uno de los países que se destaca en número de artículos indexados publicados sobre la Paratuberculosis, siendo los bovinos son la especie más estudiada, seguida por caprinos y ovinos. Así mismo existe una discreta producción de artículos enfocados en queso, leche y agua. Estos autores también observaron que las pruebas microbiológicas (cultivo), serológicas (ELISA) y moleculares (PCR) prevalecen para los siguientes propósitos: *i*) detección del agente (mayoría de los artículos); *ii*) estudios epidemiológicos; *iii*) detección de la frecuencia de anticuerpos anti MAP; *iv*) conservación de la secuencia IS*900*; v) construcción de proteína recombinante para detección de MAP y diferenciación de M. bovis; *vi*) evaluación de diferentes formulaciones de medio HEYM para cultivo de las heces; *vii*) identificación de los factores de riesgo para la enfermedad; *viii*) tipificación de cepas de MAP; *ix*) estudio de la coinfección en la glándula mamaria y; *x*) detección del fenómeno de *passthought*. Los trabajos proceden de grupos de investigación localizados prácticamente en todo el país, pero con mayor concentración en las regiones sudeste, nordeste y sur.

En cuanto al estudio de MAP en humanos, fue verificado en artículos indexados, que el país posee pocos trabajos que buscan evaluar la contribución de la bacteria en pacientes con enfermedades inflamatorias intestinales, entre ellas, la Enfermedad de Crohn. Estos trabajos tuvieron como objetivo detectar, cuantificar y evaluar los factores de riesgo con relación a la presencia de MAP en biopsias intestinales.

A pesar de que las investigaciones sobre la Paratuberculosis en Brasil han aumentado significativamente en cantidad y calidad en las últimas décadas, todavía son pocas cuando comparadas a las de otros países. Además, gran parte de los estudios nacionales presentan resultados que no buscan la continuidad de la investigación o no presentan una contribución innovadora sobre el tema abordado. La falta de una mayor integración entre los grupos de trabajo existentes, asociado a la falta de financiamiento y a la falta de incentivo del gobierno brasileño, son factores que contribuyen a este cuadro. De este modo, es evidente que los grupos que trabajan con la Paratuberculosis necesitan unirse, discutir, definir los trabajos, trazar metas y planificar un direccionamiento investigativo que consolide el tema, de igual forma que muestre la importancia de esta enfermedad en el escenario nacional y su implicación en el escenario internacional.

En Brasil no existe un Programa Nacional para esta enfermedad, pero la Paratuberculosis está presente en la lista de enfermedades de notificación inmediata y obligatoria para cualquier caso confirmado al Servicio Veterinario Oficial (IN nº 50 24/09/2013). Sin embargo, en esta IN no se define que prueba será utilizada para su confirmación. La utilización de vacunas ampliamente utilizadas en diversos países y que instituyeron el programa de control de la Paratuberculosis, todavía está prohibida en Brasil por interferir en las pruebas diagnósticas para la tuberculosis bovina y bufalina prevista en el Programa Nacional de Control y Erradicación de la Brucelosis y Tuberculosis Animal. El desconocimiento de la real importancia económica y social de la Paratuberculosis en Brasil hace que esta enfermedad sea considerada de poca importancia para que el gobierno instituya medidas de control. Las asociaciones de diferentes instituciones nacionales e internacionales son fundamentales no sólo para la elaboración de investigaciones con resultados reales y de importancia práctica, sino también para solicitar un posicionamiento más concreto del gobierno para el control de la Paratuberculosis en Brasil.
El control de la Paratuberculosis en Brasil se ve obstaculizado por las extensas áreas de bosques que albergan animales silvestres que pueden ser potenciales portadores del agente etiológico. No existen estudios indexados que investiguen la importancia de estos animales en la cadena epidemiológica de la Paratuberculosis en el territorio nacional, ni el riesgo de transmisión a los animales domésticos.

Existen aún muchos desafíos para que se logre incrementar las investigaciones y el control de la Paratuberculosis en Brasil, sin embargo, estos desafíos pueden ser superados con éxito dependiendo del interés y del esfuerzo de cada uno de los segmentos de la sociedad involucrados.

Agradecimientos
Soporte financiero: FAPEMIG (Fundação de Amparo à Pesquisa de Minas Gerais), CNPq (Conselho Nacional de Desenvolvimento Científico e Tecnológico,) y CAPES (Coordenação de Aperfeiçoamento de Pessoal de Nível Superior), Brasil.

Bibliografía

Batida, F; Juste, R.A. Paratuberculosis: a review with a focus on vaccination. Journal of Immune Based Therapies and Vaccines. 2011; 9: 8.

-Carvalho, I. A. ; Pietralonga, P. A. G. ; Schwarz, D. G. G. ; faria, A. C. S. ; Braga, I. F. E. ; Carvalho G. D. ; Valente, F. L. ; Machado, J. P. ; Guimaraes, I. M. ; Ferrari, M. L. A. ; Silva Júnior, A. ; Moreira, M. A. S. . Presence of *Mycobacterium avium* subsp. *paratuberculosis* (MAP) in Brazilian patients with inflammatory bowel diseases and in controls. Sao Paulo Medical Journal, p. 00-00, 2015.

Dacorso Filho, P.; Campos, I.O.N.; Faria, J.F.; Langenegger, J. Doença de Johne (paratuberculose) em bovinos nacionais. Arquivos do Instituto Biologia Animal, v.3, p.129-139, 1960.

-Dupont, O. Jornal do Commercio, v.5 de novembro, p.8. 1915.

-Espeschit I. F. ; Carvalho , I. A. ; Oliveira, J. F. ; Souza, M. C. C. ; Valencio, A. ; Ferrari, M. L. A. ; Moreira, M. A. S. . Risk factors for the Presence of *Mycobacterium avium* spp.*paratuberculosis* in Human Intestinal Biopsies in Brazil. Austin Journal of Gastroenterology, v. 4, p. 1-6, 2017.

-Espeschit I. F.; Schwarz1, D. G. G.; Faria, A. C. S; Souza, M. C. C., Paolicchi, F. A., Juste, R. A.; Carvalho, I. A; Moreira, M. A. S. Paratuberculosis in Latin America: a systematic review. Tropical Animal Health and Production. 49(8):1557-1576, 2017.

-Instituto Brasileiro de Pesquisa. https://brasilemsintese.ibge.gov.br/territorio/dados-geograficos.html (março de 2017).

-Ministério da Agricultura pecuária e Abastecimento. http://www.agricultura.gov.br/assuntos/sanidade-animal-e-vegetal/saude-animal/programas-de-saude-animal (março de 2017).

PASADO, PRESENTE Y FUTURO DE LA PARATUBERCULOSIS EN CHILE

Miguel Salgado
Instituto de Medicina Preventiva Veterinaria, Facultad de Ciencias Veterinarias, Universidad Austral de Chile. Valdivia, Chile. *miguelsalgado@uach.cl*

INTRODUCCIÓN

La paratuberculosis, o Enfermedad de Johne, es una enfermedad infecciosa crónica de origen bacteriano causada por *Mycobacterium avium* subsp. *paratuberculosis (M. paratuberculosis* o MAP), ampliamente diseminada en el mundo entero. A pesar de la antigüedad de su descripción, de las características manifestaciones clínicas y patológicas, y de la diversidad de especies animales susceptibles, la mayoría de los productores de ganado están poco familiarizados con esta infección, pasando muchas veces desapercibida.

Posterior a la descripción de Johne y Frothingham, la enfermedad fue reconocida y descrita en la mayoría de los países de Europa, pasando posteriormente a Australia, Estados Unidos, Africa, Asia y Sudamérica (Chiodini, 1993), debido probablemente a la comercialización de ganado infectado, una de las formas más eficientes para diseminar la infección. Actualmente la paratuberculosis es considerada una epidemia mundial que avanza lentamente como un glaciar y nuestro país no escapa a esta realidad.

PARATUBERCULOSIS EN CHILE

Aunque es probable que la paratuberculosis haya estado presente en Chile desde hace muchos años atrás, sólo fue descrita por primera vez en 1958 en el Instituto de Microbiología de la Universidad Austral de Chile por Grinbergs y Caorsi en la provincia de Valdivia, quienes estudiaron tres casos bovinos (1 toro y 2 vacas de 4 años) con diarrea crónica intermitente, logrando aislar el agente etiológico en los tres casos estudiados. Las manifestaciones clínicas y las lesiones anátomo-patológicas observadas en estos animales correspondían exactamente a la clásica descripción de la enfermedad. Posteriormente, Zamora y Reinhardt (1969), también en el Instituto de Microbiología de la Universidad Austral de Chile, obtienen un 35,9% de muestras positivas al examen microscópico directo (tinción de Ziehl-Neelsen) de materia fecal de animales con signos clínicos de paratuberculosis procedentes de las provincias de Valdivia, Osorno y Llanquihue, confirmando además la infección por exámenes histopatológicos. Estos hallazgos motivaron a este grupo de investigadores a continuar investigando esta enfermedad, particularmente mejorando las técnicas de diagnóstico bacteriológico. Es así como Faune (1970) evalúa el uso de un compuesto de amonio cuaternario (zefirán) junto con el tradicional ácido oxálico como descontaminante fecal para el aislamiento de M. paratuberculosis y compara el cultivo fecal (medio de Herrold modificado) con el examen microscópico directo y la prueba de hipersensibilidad cutánea (tuberculina aviar). De un total de 79 animales examinados provenientes de dos rebaños lecheros de las provincias de Cautín y Osorno con antecedentes clínicos de la enfermedad, se logró aislar 10 (12.6%) cepas de M. paratuberculosis de animales aparentemente sanos, concluyendo que el medio de Herrold junto con el zefirán al 0,24% como descontaminante de la muestra fecal, resultó ser más eficaz para detectar animales infectados que el examen microscópico directo y la prueba de hipersensibilidad. A pesar de estos concluyentes resultados, el examen microscópico directo de material fecal o raspado de mucosa rectal seguía siendo el método más utilizado en Chile para el diagnóstico de la enfermedad, por su rapidez, facilidad y bajo costo, a pesar de su baja sensibilidad diagnóstica. En consecuencia, algunos años más tarde Zamora y col. (1975) deciden comparar 4 procedimientos para el examen microscópico directo de material fecal para el diagnóstico de paratuberculosis: a) Cunningham-Gilmour, b) centrifugación, c) flotación, y d) convencional. De un total de 141 muestras fecales analizadas de animales provenientes de rebaños con antecedentes de paratuberculosis, 48 (34%) resultaron positivas al examen microscópico directo, alcanzando los más altos porcentajes de positividad los métodos de Cunningham-Gilmour (31,9%) y Centrifugación (30,5%). Todos estos resultados, independientemente del método de diagnóstico utilizado, permiten concluir que la paratuberculosis bovina estaba ampliamente diseminada, por lo menos, en los rebaños lecheros del sur del país.

En 1975 se registra un nuevo hecho histórico en Chile en relación a paratuberculosis, cuando Zamora y col. (1975) describen por primera vez la enfermedad en ovinos, también en la provincia de Valdivia. Mediante exámenes clínicos, histopatológicos y bacteriológicos estos autores diagnostican la enfermedad en una oveja de 5 años de edad, en avanzado estado de gestación y con signos clínicos sospechosos de paratuberculosis.

Posterior a esta fecha existen antecedentes, aunque no publicados, del diagnóstico clínico de paratuberculosis bovina en numerosas oportunidades por Médicos Veterinarios de terreno que trabajan con productores lecheros en la zona sur del país. Un análisis de los registros del Laboratorio de Diagnóstico del Instituto de Microbiología de la Universidad Austral de Chile permite comprobar que sólo entre los años 1990-2000 se recibieron 138 muestras fecales de animales sospechosos de paratuberculosis enviadas por Médicos Veterinarios privados o productores lecheros, de las cuales el 25% resultaron positivas al examen microscópico directo (tinción de Ziehl Neelsen). Considerando que esta técnica de diagnóstico tiene una baja sensibilidad para detectar infección (no superior al 15%), es posible concluir que una mayor proporción de animales estaban infectados subestimando la real magnitud del problema.

El fuerte impacto económico en animales de producción y el posible rol zoonótico de la paratuberculosis han motivado en la última década a numerosos investigadores en diferentes países del mundo, y también en Chile, a evaluar la real magnitud del problema, mejorando y desarrollando nuevas técnicas diagnósticas para detectar a los animales infectados. Lamentablemente en Chile no existen antecedentes oficiales de prevalencia a nivel nacional, pero un estudio serológico (ELISA IDEXX) realizado en 1996 por el Ministerio de Agricultura en la V, VI, VII y RM (datos no publicados) , en un universo de 84 rebaños lecheros y 1.185 vacas, demostró una seroprevalencia de 37% de rebaños positivos y una seroprevalencia individual de 2,8%, cifras muy similares a las descritas en otros países. Posteriormente, algunos laboratorios de diagnóstico bacteriológico privados, particularmente COOPRINSEM, Osorno, implementaron el diagnóstico serológico rutinario mediante el uso de la prueba de ELISA, para satisfacer la creciente demanda de los productores de la zona sur del país. A manera de ejemplo, un análisis de los resultados obtenidos en el Laboratorio de Bacteriología de COOPRINSEM sólo durante el año 2000 se analizaron 693 muestras de animales provenientes de 23 rebaños lecheros de la zona sur del país, de los cuales un 52% de los rebaños y un 16% de los animales resultaron reaccionantes a la prueba de ELISA (datos no publicados) . Obviamente, estos resultados no pueden considerarse como prevalencia puesto que corresponden a una muestra selectiva de rebaños con sospecha clínica de paratuberculosis, pero sin duda demuestran el interés de los productores lecheros por conocer el estatus de infección de sus rebaños y el alto grado de diseminación de la enfermedad.

El interés y la preocupación por esta enfermedad re-emergente en Chile se traspasaron rápidamente a las Universidades y es así como a partir del año 2000 se inician una serie de estudios, especialmente en las Facultades de Medicina Veterinaria de las Universidades de Chile y Austral de Chile, aportando nuevos e importantes antecedentes sobre la paratuberculosis en Chile. Entre estas investigaciones es posible destacar las siguientes.

a) En el año 2000, en el Instituto de Microbiología de la Universidad Austral de Chile se implementa y estandariza oficialmente el método de diagnóstico por cultivo fecal en medios sólidos (medio de Herrold modificado) siguiendo las recomendaciones y protocolos desarrollados en la Universidad de Cornell, USA y en el USDA, USA (Soto y col., 2002a). Simultáneamente se realiza un estudio para evaluar la sensibilidad y especificidad de las dos pruebas serológicas de ELISA (IDEXX y SVANOVA) que se estaban utilizando en Chile para el diagnóstico de paratuberculosis. Con este propósito se analizaron muestras séricas y fecales de 250 animales provenientes de 14 rebaños de la X Región con antecedentes clínicos de la enfermedad. En el 71.4% de los rebaños y 16% de los animales fue posible aislar M.paratuberculosis; el 64% de los rebaños y 8% de los animales resultaron reaccionantes a la prueba de ELISA IDEXX, determinándose una sensibilidad y especificidad para esta prueba de 32.5% y 96.7%, respectivamente. Por el contrario, la prueba de ELISA SVANOVA resultó con una baja especificidad (86%), presentando un elevado porcentaje de resultados falsos positivos, concluyéndose que su uso no sería recomendable para el diagnóstico de paratuberculosis en Chile (Soto y col., 2002b).

c) El año 2002 Ábalos y col., de la Facultad de Ciencias Veterinarias y Pecuarias de la Universidad de Chile, informan el primer aislamiento de M. paratuberculosis a partir de muestras fecales de cabras lecheras de la IV Región. Lamentablemente, este hallazgo no fue publicado y no tiene validez oficial.

b) El año 2002 Pinto y col., en la Facultad de Ciencias Veterinarias de la Universidad Santo Tomás, realiza un estudio de seroprevalencia también en rebaños caprinos de leche con sistemas de manejo intensivo de producción, determinándose una seroprevalencia individual de 9,3% sobre un universo de 2.500 cabras examinadas En este estudio no se realizó examen bacteriológico y tampoco ha sido oficialmente publicado.

d) En la Universidad Austral de Chile, Instituto de Microbiología, el año 2003 se inician una serie de estudios conducentes a evaluar diferentes técnicas de diagnóstico de paratuberculosis bovina. Gracias al apoyo financiero del Fondo de Mejoramiento del Patrimonio Sanitario del Servicio Agrícola y Ganadero, Ministerio de Agricultura, se realiza un proyecto a gran escala para validar la prueba serológica de ELISA IDEXX en una muestra estadísticamente representativa de la X Región. Se utilizaron 28 rebaños lecheros de diferentes tamaño y un total de 1.381 animales >2 años. Los resultados obtenidos demostraron una seroprevalencia de 78% a nivel de rebaño y 7% a nivel individual, aislándose el agente etiológico en el 35.7% de los rebaños y 3.9% de los animales (Kruze y col., 2005; van Schaik y col, 2006). Este mismo grupo de investigadores evalúan, además, el uso del cultivo de pool fecal para el diagnóstico bacteriológico de rebaños infectados utilizando un total de 598 muestras fecales de animales procedentes de 12 rebaños infectados con M.paratuberculosis, concluyendo que el pool fecal de 5 animales no altera la sensibilidad del cultivo fecal individual pero disminuye notablemente los costos del examen, facilitando su uso masivo en programas de control (Pradenas y col., 2004; van Schaik y col., 2007).

e) El 2006 Kruze y col. publican oficialmente por primera vez en Chile el aislamiento de M. paratuberculosis a partir de una cabra lechera con signos clínicos de la enfermedad en un rebaño lechero de la comuna de Purranque, provincia de Osorno y describen las características histopatológicas de la enfermedad. Posteriormente, estos mismos autores realizan un estudio serológico y bacteriológico en 8 rebaños caprinos de leche de diferentes regiones del país con diferentes sistemas de manejo (intensivo y extensivo). Los resultados obtenidos revelaron una alta prevalencia de infección especialmente en los rebaños con manejo intensivo, aislándose M. paratuberculosis en el 50% (4) de los rebaños y 9.1% (35) de los animales examinados (Kruze y col., 2007). Paralelamente, este mismo grupo de investigación evalúa el uso de tres kits comerciales de ELISA-bovino (IDEXX, PourquierID-Vet) para el diagnóstico de paratuberculosis en caprinos lecheros demostrándose que todos son igualmente sensibles para detectar animales infectados no existiendo diferencias significativas entre ellos (Badilla,2006). Continuando los estudios con rebaños caprinos, Salgado y col. (2007) evalúan una prueba de ELISA (IDEXX) con muestras de suero sanguíneo y leche para el diagnóstico de paratuberculosis en cabras lecheras utilizando como método de referencia el cultivo fecal. Estos autores concluyen que la prueba de ELISA con muestras de leche tiene una sensibilidad similar a la prueba de ELISA con muestras de suero para detectar animales infectados, con lo cual es posible recomendar su uso en forma masiva en programas de control.

f) La baja sensibilidad de los ELISAs comerciales disponibles actualmente en el mercado internacional motivaron al grupo de investigación de la Universidad Austral de Chile a intentar desarrollar un nuevo kit de ELISA para el diagnóstico de paratuberculosis bovina, utilizando antígenos semi-purificados extraídos de cepas nativas de M.paratuberculosis. Con esta finalidad se postuló a un proyecto FONDECY (Kruze, 2005), que se encuentra actualmente en ejecución y cuyos resultados hasta la fecha demuestran que M. paratuberculosis secreta en medios de cultivo líquidos varias proteínas inmunogénicas que podrían ser buenas candidatas para ser utilizadas como antígeno de fase sólida en una prueba de ELISA, mejorando la sensibilidad de la prueba (Jara, 2007).

g) Finalmente, es importante mencionar que la paratuberculosis en Chile no está confinada sólo a los animales rumiantes domésticos como son el bovino, ovino y caprino. El año 2007, el grupo de investigación de la Universidad Austral de Chile demostró que la enfermedad también está presente en animales de vida silvestre como guanacos y ciervos, resultados que fueron presentados recientemente en el 9º Coloquio Internacional de Paratuberculosis realizado en Tsukuba, Japón (Salgado y col., 2007a,b; Paredes y col., 2007), y también en este mismo congreso (Salgado y Kruze, 2007). Recientemente, se ha descrito la infección también en guanacos (Lama guanicoe) (Salgado y col. 2009), liebre europea (Lepus europaeus) (Salgado y col., 2011), ciervos (Pradenas y col., 2014), pudúes (Pudu puda) (Salgado y col., 2015) y alpacas (Lama pacos) (Salgado y col., 2016).

En cuanto a la prevalencia de la infección por MAP en los bovinos, un estudio publicado en los últimos años estimó que entre un 44 a un 87% de los rebaños lecheros del sur del territorio nacional, donde se concentra la mayor masa ganadera, se encontraban infectados con MAP (Kruze y col., 2013). Un estudio reciente, realizado también en el mismo sector lechero de Chile, estimó que la prevalencia verdadera a nivel individual se encontraba entre un 3 a un 25%, con una mediana de un 9% (Verdugo y col., 2016).

Colombia

PARATUBERCULOSIS IN COLOMBIA: PAST, PRESENT AND FUTURE

Jorge A Fernández-Silva, Nathalia M Correa-Valencia

Grupo Centauro, Escuela de Medicina Veterinaria, Facultad de Ciencias Agrarias, Universidad de Antioquia, Calle 70 No. 52-21, Medellín, Colombia

Colombian studies on paratuberculosis

Twenty-six original studies referring to Johne´s disease and *Mycobacterium avium* subsp. *paratuberculosis* (MAP) detection have been carried out in Colombia. So far, no study in Colombia has attempted the detection of MAP in food or humans. In addition to the 26 original studies mentioned above, 14 reviews, case reports, case series reports, and editorials were considered in this document, but they are of great value for the national knowledge on MAP or Johne´s disease (JD) and demonstrate the national concern about MAP and its impacts in Colombia through several decades. According to an unavailable document by Plata, 1931, the existence of MAP in Colombia was first documented by the Cuban veterinarian Ildefonso Pérez Vigueras in 1924, in a herd of imported cattle of the municipality of Usme (province of Cundinamarca) in cattle with Johne´s disease (Vega, 1947).

Most studies on MAP or JD have been carried out during the present decade. Most studies have been carried out in animals of the Provinces of Antioquia and Cundinamarca, some in Caldas and Tolima, and few in Nariño and Boyacá. The original studies on MAP in Colombia have reported results from cattle, sheep, goats, and buffaloes. Studies on cattle have been the most common compared to sheep and goats, and buffaloes. Other relevant species in the country (wild mammals or humans) have not been found or cited in any original study reviewed. The most common diagnostic test used to investigate MAP in Colombia is the enzyme-linked immune-assay (ELISA), followed by microscopy on Ziehl-Neelsen (ZN)-stained samples (on feces, rectal mucosa scrapings, or tissues), polymerase chain reaction (PCR), intradermal Johnin test (IJT, with bovine and/or avian-purified protein derived), culture (from feces or tissues, and individual or pooled), complement fixation, indirect immuno-fluorescence, and counter immuno-electrophoresis. The studies published so far include cross-sectional, diagnostic test comparisons, risk factor analyses, and clinical trials (on treatments). Thus far, no cohort or case and control studies have been published in Colombia.

What Colombian studies on paratuberculosis are telling us?

According to several anecdotal reports and opinions, the national or regional distribution of MAP or JD in cattle and small ruminants in Colombia is not homogeneous or conclusive. Some academics and producers consider JD as a significant problem, while others claim the absence or very low prevalence of MAP in farmed animals. The number of publications reporting original studies on MAP, especially JD, in recent years is relatively low compared to other countries in Latin America (Fernández *et al.,* 2014), but higher than expected for Colombian conditions. The increasing number of publications suggests a growing interest on MAP research in the country, as well as an increasing concern about this microorganism and its negative effects on animal health, animal production, and its zoonotic potential (public health impact) from the academic and producers. Although JD is a notifiable disease in Colombia (ICA, 2015) it is not of major concern to animal health authorities and its control is a responsibility of the farmer (Anonymous, 2010a; Fedegán, 2010; Fernández *et al.,* 2014). This could explain the low number of initiatives for the research, prevention, and control in animals, as well as for the detection of the microorganism in food, the environment, and humans.

The locations of most Colombian studies do not follow a clear trend but could be related to the high concentrations of cattle in some of the provinces (i. e. Antioquia and Cundinamarca; ICA, 2017), or to the interests of academics, scientists, or cattle producers. Since the first report in 1924, Cundinamarca has been a province with common reports of JD (Vega, 1947; Huber, 1954; Isaza, 1978; Mogollón *et al.,* 1983; Góngora and Perea, 1984; Mancipe *et al.,* 2009). This could be explained by the long tradition of the Facultad de Medicina Veterinaria of the Universidad Nacional de Colombia in Bogotá —the oldest faculty of veterinary medicine in the country, where the first studies in the early 20th century were carried out. More recently, the Province of Antioquia has been publishing the majority of original studies, all of them from academics at the Universidad de Antioquia and the Universidad CES. As expected, studies on cattle were the most common, most likely due to the size of the population in the country and to the production systems related to milk and meat. In contrast, studies on sheep populations are less common in the country and could be due to their smaller populations (ICA, 2017).

The common use of ELISA, ZN-staining, IJT is not surprising given their relatively low cost and availability of materials, qualified personnel, and infrastructure for these types of tests. However, the use of culture and PCR is becoming more common and could be related to the recent development of the diagnostic capacity in universities, compared to national laboratories and to the expansion of the reagents and equipment supplies for such diagnoses in the country. The absence of cohort and case-control studies is common in animal health research in Colombia. These high-profile observational studies, as well as the experimental approaches are more complex, laborious, demanding, and expensive, given the microbiological and pathophysiological characteristics of MAP. Nevertheless, the current MAP situation in Colombia demands additional observational studies in addition to surveys and case reports to enhance our comprehension of the epidemiological situation and to assess the true zoonotic threat.

Definitively, the country needs to cover some knowledge gaps to get to a true understanding of the disease. It is necessary to define the exact status of the disease through well-designed prevalence/incidence studies, considering that no whole national data is available. On this refer, just some local estimates are available so far (Patiño and Estrada, 1999; Ramírez *et al.,* 2001; Fernández *et al.,* 2011a; 2011b; Benavides *et al.,* 2016; Correa *et al.,* 2016). The harmonization of diagnostic methods, considering the epidemiologic and biological behavior of MAP under local agro-ecological, productive, and cultural conditions is also needed. In addition, the laboratory infrastructure — mainly developed for foot-and-mouth disease control, should cover other entities with relevance for public health and international trade such as JD (Calderón and Góngora, 2008), improving their testing capacity and the access to diagnostic reagents. It is also necessary to improve the training of the farmers, bringing them closer to the importance of disease control, not only of Johne´s disease, but also of many others that generate economic losses and are considered of sanitary risk.

Only one study reported the molecular characterization of strains isolated in Colombian territories (Fernández *et al.,* 2011b), being this insufficient to consider the definition of "indigenous strains" and the ulterior design of vaccines. It would be necessary to carry on studies on wider regions, considering infection-assessment on cattle and other-than-cattle susceptible populations (even local wildlife), and, in this way, generate our own prophylactic strategies, according to Colombian

MAP molecular and epidemiological diversity. The relationship between MAP and Crohn´s disease (CD) has been essentially not discussed in academic fields in the country, except for some sporadic reviews (Góngora and Villamil, 1999; Calderón and Góngora, 2008; de Waard, 2010). The CD has been known in Colombia since the 1950s and the incidence and prevalence rates are increasing (estimated point prevalence of 77,000 CD cases), but no national consolidated information about the disease is available (Calderón and Góngora, 2008). According to some of these authors, efforts should be made to correlate these two diseases in areas with a high prevalence or incidence of both. In general, important progress has been made on MAP research in the areas of diagnosis and epidemiology as is reported by the studies included in this report. However, many unanswered questions remain and research opportunities in the country are plentiful.

References

Anonymous. Situación en Colombia de enfermedades bovinas no sujetas al control oficial. Fedegan. Primera edición. Bogotá, Colombia: Sanmartín Obregón & Cia; 2010a. 126 p.

Benavides B, Arteaga A, Montezuma CA. Estudio epidemiológico de paratuberculosis bovina en hatos lecheros del sur de Nariño, Colombia. Rev Med Vet 2016; 31:57-66.

Calderón J, Góngora A. Similaridades clinopatológicas entre paratuberculosis y enfermedad de Crohn ¿posible vínculo zoonótico? Rev MVZ Córdoba 2008; 13(1):1226-1239.

Correa NM, Ramírez NF, Olivera M, Fernández JA. Milk yield and lactation stage are associated with positive results to ELISA for *Mycobacterium avium* subsp. *paratuberculosis* in dairy cows from Northern Antioquia, Colombia: A preliminary study. Trop Anim Health Prod 2016; 48(6):1191-1200.

Waard JH. ¿Ordeñando micobacterias del ganado? Impacto económico y en salud de tuberculosis bovina y paratuberculosis en Colombia. Rev MVZ Córdoba 2010; 15(2):2037-2040.

Fedegán (Federación colombiana de ganaderos). Situación en Colombia de enfermedades bovinas no sujetas al control oficial. Primera edición. Bogotá, Colombia: 2010. 118 p.

Fernández JA, Abdulmawjood A, Akineden O, Bulte M. Serological and molecular detection of *Mycobacterium avium* subsp. *paratuberculosis* in cattle of dairy herds in Colombia. Trop Anim Health Prod 2011a; 43(8):1501-1507.

Fernández JA, Abdulmawjood A, Bulte M. Diagnosis and molecular characterization of *Mycobacterium avium* subsp. *paratuberculosis* from dairy cows in Colombia. Vet Med Int 2011b; 352561.

Fernández JA, Correa NM, Ramírez N. Systematic review of the prevalence of paratuberculosis in cattle, sheep, and goats in Latin America and the Caribbean. Trop Anim Health Prod 2014; 46(8):1321-1340.

Góngora OA, Perea J. Evaluación de tres métodos diagnósticos en paratuberculosis bovina. [Tesis]. Bogotá, Colombia. UNAL; 1984.

Góngora OA, Villamil JC. La paratuberculosis bovina desde la óptica de la salud pública. Holstein Colomb 1999; 147:44-48.

Huber G. La administración de la Isonicotimilhidrazina de cortisona en la paratuberculosis bovina (enfermedad de Johne). UNAL. 1954.

ICA (Instituto Colombiano Agropecuario). Mapa de distribución de bovinos y ovinos 2017 URL: https://www.ica.gov.co/Areas/Pecuaria/Servicios/Epidemiologia-Veterinaria/Censos-2016/Censo-2017.aspx

ICA (Instituto Colombiano Agropecuario) Resolución 0003714 de 2015.

Isaza PF. Diagnóstico de paratuberculosis en bovinos por los métodos de baciloscopia, fijación de complemento e inmunofluorescencia. UNAL; 1978.

Mancipe LF, Sánchez L, Rodríguez G. Estudio de la paratuberculosis en un rebaño de ovinos de la Sabana de Bogotá mediante la utilización de tres técnicas diagnósticas. Rev Med Vet 2009; 18:33-51.

Mogollón JD, Hernández AL, Tovar AL, Murillo BN, Peña NE, Mossos NA. Prevalencia de paratuberculosis ovina en el altiplano cundi-boyacense. Revista ICA (Colombia) 1983; 18:479-484.

Patiño DA, Estrada M. Determinación de la prevalencia de paratuberculosis en tres hatos del Páramo de Letras. [Thesis]. Caldas, Colombia. Universidad de Caldas; 1999.

Ramírez N, Gaviria G, Restrepo LF, Gómez C. Diagnóstico epidemiológico referente a varias patologías de bovinos en tres haciendas de la Universidad de Antioquia. (Unpublished document), 2001.

Vega A. Relación entre el diagnóstico de la paratuberculosis bovina por el examen coprológico y de la prueba alérgica de termorreacción con la tuberculina aviaria por vía subcutánea [Thesis]. Bogotá, Colombia. UNAL; 1947.

LA PARATUBERCULOSIS EN ESPAÑA, ABORDAJE DEL CONTROL DE LA ENFERMEDAD Y PRINCIPALES LÍNEAS DE INVESTIGACIÓN

Joseba M. Garrido[1], Ramón Juste[2], Valentín Pérez[3]

1. NEIKER-Instituto Vasco de Investigación y Desarrollo Agrario. Departamento de Sanidad Animal, Derio, Bizkaia, Spain.
2. SERIDA, Servicio Regional de Investigación y Desarrollo Agroalimentario de Asturias, Villaviciosa, Asturias, Spain.
3. Departamento de Sanidad Animal. Facultad de Veterinaria, Universidad de León. León, Spain.

SITUACIÓN DE LA PARATUBERCULOSIS

La paratuberculosis en España, al igual que en la gran mayoría de los países, se encuentra ampliamente extendida y afecta al ganado bovino, ovino y caprino principalmente. No hay estudios recientes que digan con precisión la prevalencia actual, pero Vázquez y cols. [1] encontraron una prevalencia individual en ganado bovino lechero del 46.7% en base a las lesiones histopatológicas y del 39,1% en base a PCR de tejidos. En un estudio similar Balseiro y cols. [2] observaron una prevalencia del 28,4%. Respecto a la prevalencia a nivel de rebaño, se ha estimado que en algunas regiones de España el 50% de las explotaciones de ganado bovino tienen al menos un animal infectado [3]. Respecto a las especies ovina y caprina los estudios de prevalencia son anteriores y se habla de una prevalencia a nivel de rebaño en la especie ovina en la Comunidad de Aragón del 46,7% en base a una combinación de técnicas microbiológicas, serológicas y anatomopatológicas [4 y del 31,4% en el País Vasco en base a la técnica ELISA en un estudio llevado a cabo sobre 226 rebaños por Aduriz. [5]. En este último estudio la prevalencia individual se estimó en un 5,8%. En estudios de matadero llevados a cabo en la provincia de León por González y cols. [6] observaron lesiones compatibles con la enfermedad en el 24,3% de los animales analizados. Finalmente, los resultados obtenidos en estudios llevados a cabo en la especie caprina en la década de los noventa, presentan aún una mayor variabilidad. Mediante la técnica ELISA, utilizando diferentes antígenos, se observaron prevalencias individuales del 41%, 13% y 0,5% en las provincias de Huelva, Córdoba y en la Comunidad Canaria, respectivamente según Molina y cols. [7]. Posteriormente, en un estudio llevado a cabo por Rodríguez [8] en la provincia de Córdoba se llegó a considerar que el 97% de las explotaciones caprinas estaban afectadas y que la prevalencia individual era del 49,7%.

Esta situación, excepto en la especie ovina donde el control mediante vacunación en las explotaciones afectadas es una práctica generalizada, parece haberse mantenido en el tiempo. En estos momentos la paratuberculosis es considerada como una de las enfermedades que causa mayores pérdidas económicas en las explotaciones de ganado bovino y principalmente en el ganado bovino lechero.

En España no existe un programa nacional de control de la paratuberculosis, pero hay Comunidades Autónomas o regiones en las que existen programas voluntarios para el control de la enfermedad.

CONTROL DE LA PARATUBERCULOSIS

Tal y como se ha mencionado anteriormente el control de la paratuberculosis en España se gestiona de forma particular o a través de programas voluntarios en algunas Comunidades Autónomas. A nivel general los programas de control planteados se basan en la mejora del manejo y en la detección y eliminación de los animales positivos. En estos programas la técnica de elección, basándose en aspectos económicos y de facilidad de procesamiento de grandes cantidades de muestras, es el ELISA indirecto para la detección de anticuerpos. Sin embargo, lo más importante a nivel epidemiológico es la detección de los animales excretores, por lo que siempre es recomendable la detección de Map en heces mediante PCR con el fin de reducir el número de animales a eliminar. Añadir esta técnica encarece los programas de control por lo que es algo que debe considerarse en base a la relación coste-beneficio, ya que habrá animales no excretores que podrían continuar su vida productiva, al menos, hasta el final de la lactación. En cualquier caso, cada vez son mayores las evidencias que indican que la mejor estrategia de control en base a la relación coste-beneficio es la vacunación. La vacunación frente a paratuberculosis es el método de control de elección en el caso de la especie ovina y los resultados han sido óptimos. Sin embargo, la vacunación frente a paratuberculosis en ganado bovino no está permitida en España por la posible interferencia con el diagnóstico de la Tuberculosis bovina, enfermedad sometida a Campañas Oficiales, cuya erradicación se considera prioritaria. En el País Vasco, a día de hoy, la situación frente a la tuberculosis es muy buena, con un 0,09% de las explotaciones afectadas en 2017. Esto ha hecho que desde 2005, NEIKER junto con los Servicios de Ganadería de las Diputaciones Forales, estemos llevando a cabo un ensayo de campo con una vacuna inactivada frente a paratuberculosis con el permiso de la Agencia Española de Medicamentos y Productos Sanitarios y del Ministerio de Agricultura, Pesca, Alimentación y Medio Ambiente. El estudio empezó en 2005 y en estos momentos están participando 20 explotaciones que siguen un programa de vacunación y nueve que siguen otro de detección y eliminación de positivos. En un primer momento se vacunaron todos los animales presentes en la explotación y posteriormente la reposición con una vacuna inactivada (Silirum®, CZ Veterinaria, España). En todos ellos se hacen muestreos anuales de sangre para la detección de anticuerpos mediante ELISA y de heces para la detección de Map mediante PCR a tiempo real de los animales mayores de 24 meses, así como la prueba de la Intradermorreacción comparativa con PPD aviar y PPD bovina para el diagnóstico de la tuberculosis en todos los animales mayores de 45 días. Los resultados obtenidos con ambas estrategias de control se consideraron satisfactorios, pero el coste económico soportado por las explotaciones control fue muy superior, ya que en estas se requería el sacrificio de los animales con resultado positivo en los análisis laboratoriales, mientras que los que fueron incluidos en el programa de vacunación no fue necesario eliminar ningún animal por esta razón. Entre las 20 explotaciones vacunadas hay seis en las que no se han encontrado animales excretores de Map con las heces en los últimos seis muestreos anuales, por lo que se puede considerar que Map ha sido erradicada en estas explotaciones

PRINCIPALES LÍNEAS DE INVESTIGACIÓN

Diagnóstico la paratuberculosis bovina
Estudios de interferencia de la vacunación frente a Map con el diagnóstico de la tuberculosis
Desarrollo de técnicas moleculares, microbiológicas y serológicas
Control de la paratuberculosis bovina
Ensayos de vacunación en campo
Desarrollo de nuevos productos vacunales que no interfieran en el diagnóstico serológico de la paratuberculosis ni con las técnicas oficiales de diagnóstico de la tuberculosis bovina
Verificación de marcadores genéticos para selección de bovinos resistentes a la paratuberculosis
Estudios de patogenia y respuesta inmune asociados a la infección y vacunación frente a paratuberculosis en ovino y caprino
In vivo. Modelo de infección de pequeño rumiante en cabra
Ex vivo. Modelo de infección macrófagos y neutrófilos caprinos
Identificación de marcadores de protección asociados a la vacunación
Inmunofenotipado de lesiones en ovino

Desarrollo de modelos con el fin de realizar ensayos rápidos y económicos de patogenia, cribado de nuevas terapias e interferencia con el diagnóstico de la tuberculosis

In vivo. Modelo de infección leporino

In vivo. Modelo de interferencia cobaya

Ex vivo. Modelo de infección macrófagos de rumiantes y neutrófilos de rumiantes y conejo

Existen colaboraciones activas entre los grupos de micobacterias de diferentes instituciones lo que permite llevar a cabo proyectos de investigación multidisciplinares gracias a la complementariedad de su conocimiento.

REFERENCIAS

Vazquez P, Garrido JM, Molina E, Geijo MV, Gomez N, Perez V, Sevilla IA, Alonso-HeErn M, Cortes A, Juste RA (2014). Latent infections are the most frequent form of paratuberculosis in slaughtered Friesian cattle. Span. J. Agric. Res. 12(4):1049–1060.

Balseiro A, Prieto JM, Espi A, Perez V, Garcia Marin JF. (2003). Presence of focal and multifocal paratuberculosis lesions in mesenteric lymph nodes and the ileocaecal valve of cattle positive to the tuberculin skin test. Vet. J. 166(2):210–212.

Perez V, Moreno O, Muñoz M, Garcia-Pariente C, Benavides J, Delgado L, et al. (2009). Diagnosis of paratuberculosis in slaughtered calves in the Northwest of Castilla y León (Spain) by pathological methods. In: Proceedings of the 10th international colloquium on paratuberculosis (10ICP 2009), Minneapolis, 9–14 Aug 2009, p 184.

Juste RA, Badiola JJ, Amal MC, Balaguer L, García Marín JF, Sáez de Ocáriz C and Marco JC. A survey of ovine paratuberculosis in Aragón (Spain) by different methods. The Paratuberculosis Newsletter, 3(2): 3-4. 1991.

Aduriz G. (1993). Epidemiología, diagnóstico y control de la paratuberculosis ovina en la Comunidad Autónoma del País Vasco. Tesis Doctoral. Universidad de Zaragoza.

González J. (1998). Empleo de un método histológico para el estudio de la prevalencia de la paratuberculosis ovina. Tesina de Licenciatura. Universidad de León.

Molina A, Morera L y Llanes D. (1991). Enzyme-linked immunosorbent assay for detection of antibodies against *Mycobacterium paratuberculosis* in goats. Am J Vet Res, 52(6):863-868.

Rodríguez S. (2000). Desarrollo y evaluación de una prueba ELISA indirecta para el serodiagnóstico de paratuberculosis ovina. Tesina de Licenciatura. Universidad de Córdoba.

PARATUBERCULOSIS EN MÉXICO, AVANCES, PERSPECTIVAS Y RETOS

Gilberto Chávez Gris[1], Edith Maldonado Castro[1], Ana Laura Hernández Reyes[1], Victoria Castrellón Ahumada[1]

[1]Universidad Nacional autónoma de México. Facultad de Medicina Veterinaria y Zootecnia de la UNAM. Centro de Enseñanza, Investigación y Extensión en Producción Animal en Altiplano (CEIEPAA), Tequisquiapan, Querétaro, México.
gris@unam.mx

La paratuberculosis en México ha sido diagnosticada desde hace más de 50 años en México y el primer aislamiento se realizó en la década de los 70's. Actualmente se considera una enfermedad que está presente en el país y en diferentes especies de rumiantes. Podemos decir que en aquellos estados donde se ha realizado su diagnóstico se ha encontrado a esta micobacteriosis, como se muestra en la siguiente figura.

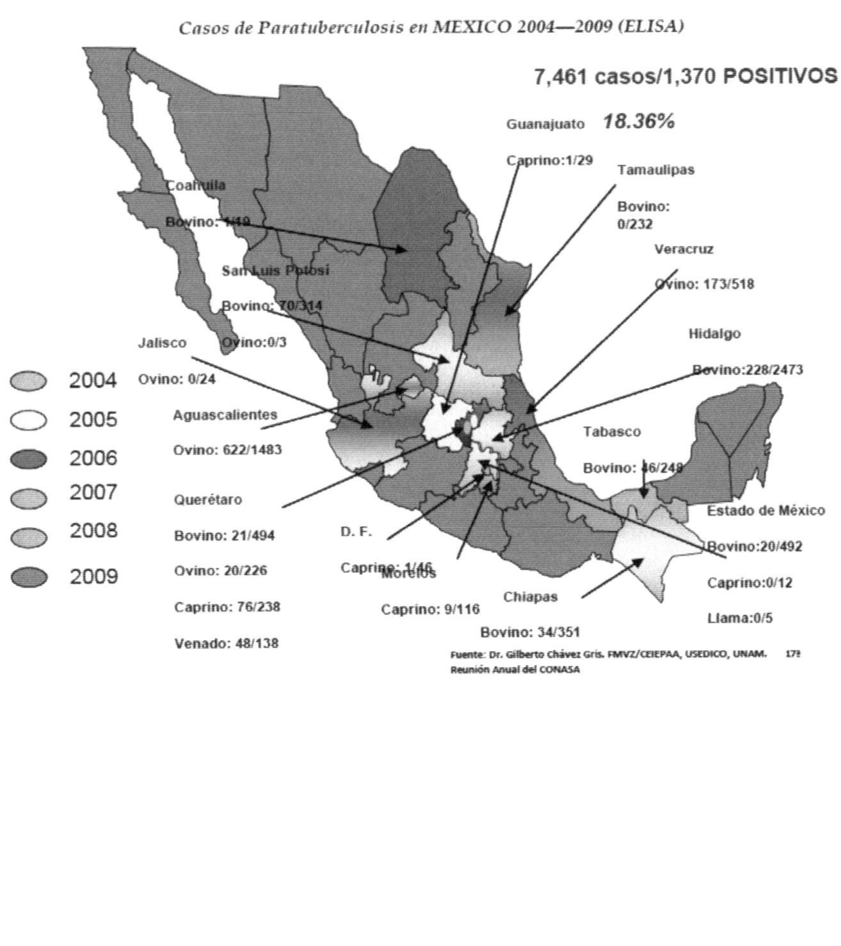

En México no existe una campaña para el control de la Paratuberculosis, como si ocurre para Brucelosis o Tuberculosis; esto debido a que no se ha considerado como de un impacto relevante y que tenga un impacto sobre la Salud Pública. Sin embargo, existen estudios en México realizados sobre el impacto económico que genera la paratuberculosis en bovino lechero en una población de casi 30,000 animales observándose que con una prevalencia del 8,87% el impacto se calculó en más de $10,000 (pesos mexicanos). Así también se realizó en estudio en tejidos de pacientes mexicanos diagnosticados con la enfermedad de Crohn y mediante Hibridación In Situ se detectó positividad sobre todo en aquellos casos que mostraban lesiones granulomatosas. Se han realizado estudios sobre las diferencias genéticas en los aislados de los bovinos, ovinos, caprinos y animales de zoológico identificándose cepas tipo C, S e I, siendo la más frecuente la del tipo C.

En 2010, se desarrolló una propuesta por un grupo de trabajo del CONASA (Consejo Técnico Consultivo Nacional) denominado "Plan estratégico del Programa para la atención de la Paratuberculosis en ganado bovino, ovino y caprino en México", que considera varios aspectos a desarrollar como son: 1) El fomento del conocimiento de la Enfermedad entre productores y Veterinarios, 2) Desarrollar un Programa voluntario de prevención y control, 3) Control de movilizaciones y 4) Operación de un Sistema de Vigilancia.

Entre los puntos relevantes de la paratuberculosis se consideró lo siguiente:

1) Pérdidas económicas. Esta dependerá del sistema de producción y de la prevalencia que se tenga y puede ir desde pérdidas incipientes hasta pérdidas relevantes. Sin embargo, esto no significa que en todos los casos será igual, por lo que cada situación sanitaria será diferente.
2) Disminución en la producción láctea. Esta disminución en la reducción ha sido evaluada como una de las más importantes en el impacto económico que tiene la Ptb en la ganadería especializada en producción lechera y dependiendo de su prevalencia las pérdidas pueden llegar a ser relevantes.
3) Desecho de animales. En aquellos hatos o rebaños con elevada prevalencia, se ha observado un aumento y precocidad en el desecho de los animales afectados, con la consecuente pérdida del material genético.
4) Restricción en la venta de pie de cría. Eventualmente la venta de estos ejemplares se puede ver afectada, cuando se les solicita alguna constancia o prueba de ser negativo o pertenecer a un hato o rebaño libre a Ptb, siendo una limitante eventualmente para su comercio.
5) Posible zoonosis. También por su posible asociación como zoonosis con la Enfermedad de Crohn (EC). Existen diversos trabajos que muestran evidencias de su probable asociación, sin embargo, también algunos trabajos no han demostrado la participación de Map en la EC.

En México, se ha observado la presencia de Map en diversos estados del país, aunque no se tiene un estudio epidemiológico se sabe que está presente y afecta a ovinos, caprinos, bovino productor de leche, bovino productor de carne, ganado de lidia, siendo este sector uno de los más sensibilizados ante la paratuberculosis. Asimismo, en rumiantes no domésticos como los de Zoológico, se ha encontrado la presencia de Map y llega a generar importantes pérdidas de animales. Se ha logrado la identificación de los diferentes aislamientos de Map. Con base en los antecedentes en nuestro país, se ha recomendado llevar a cabo estrategias tanto para el diagnóstico como para su control.

Para que se pueda establecer el control de la Paratuberculosis en México con probabilidades de éxito, se requiere de varios considerar los siguientes aspectos:

Conocer la biología de la paratuberculosis. Esto debe considerar la educación a productores y eventualmente a veterinarios clínicos. Esta educación es determinante para que se comprendan los mecanismos de transmisión.
Convencimiento por parte de productores. Este punto está ligado al anterior ya que solamente cuando se conoce la enfermedad y los posibles impactos sanitarios y económicos se puede tomar la mejor decisión. Así deben contemplarse las condiciones y características de su sistema de producción para elaborar una estrategia de control con el convencimiento de que tendrá un efecto positivo en la situación sanitaria y sobre la producción.
Contar con laboratorios que realicen diagnóstico de Ptb. Debido a que los signos clínicos solo se presentarán en una pequeña proporción de los animales infectados es necesario realizar el diagnóstico a través de pruebas serológicas para conocer su prevalencia, así como de aislamiento bacteriano, pruebas moleculares como PCR y estudios anatomopatológicos, esta última en la mortalidad o animales con pobre condición corporal para confirmar el diagnóstico. Sin embargo, es necesario contar con laboratorios capacitados y que ofrezcan a las pruebas de diagnóstico, pero también es importante saber interpretarlas. En México, existe una pobre cobertura de laboratorios que realicen el diagnóstico de paratuberculosis para la potencial demanda Nacional.
Estrategia de Control a seguir. Con base en los resultados de laboratorio, deben plantearse las posibles estrategias para el control del hato, ya que no pueden darse lineamientos absolutos para el control, debido a las particularidades de cada hato o rebaño y debe ser construida idealmente en conjunto con el productor y el veterinario responsable del hato o rebaño.

Un programa de Control es una inversión. Considerar que toda estrategia de control requiere de una inversión y eventualmente cambios en el manejo, que además impactará de manera positiva en el control de otros procesos infecciosos.

Actualmente, en México se está trabajando en diferentes puntos del país en diagnóstico buscando contar con pruebas más accesibles para los productores, desarrollo de inmunógenos para el control de la paratuberculosis. Existen pocos laboratorios que ofrezcan el diagnóstico de la paratuberculosis en el país que no supera el número de diez en todo el territorio nacional.

Consideramos debe contemplarse una estrategia Nacional en México a través de un Programa Voluntario, que permita avanzar paulatinamente en la mejora de la situación sanitaria de manera paulatina según la capacidad de los productores.

Author index

Map

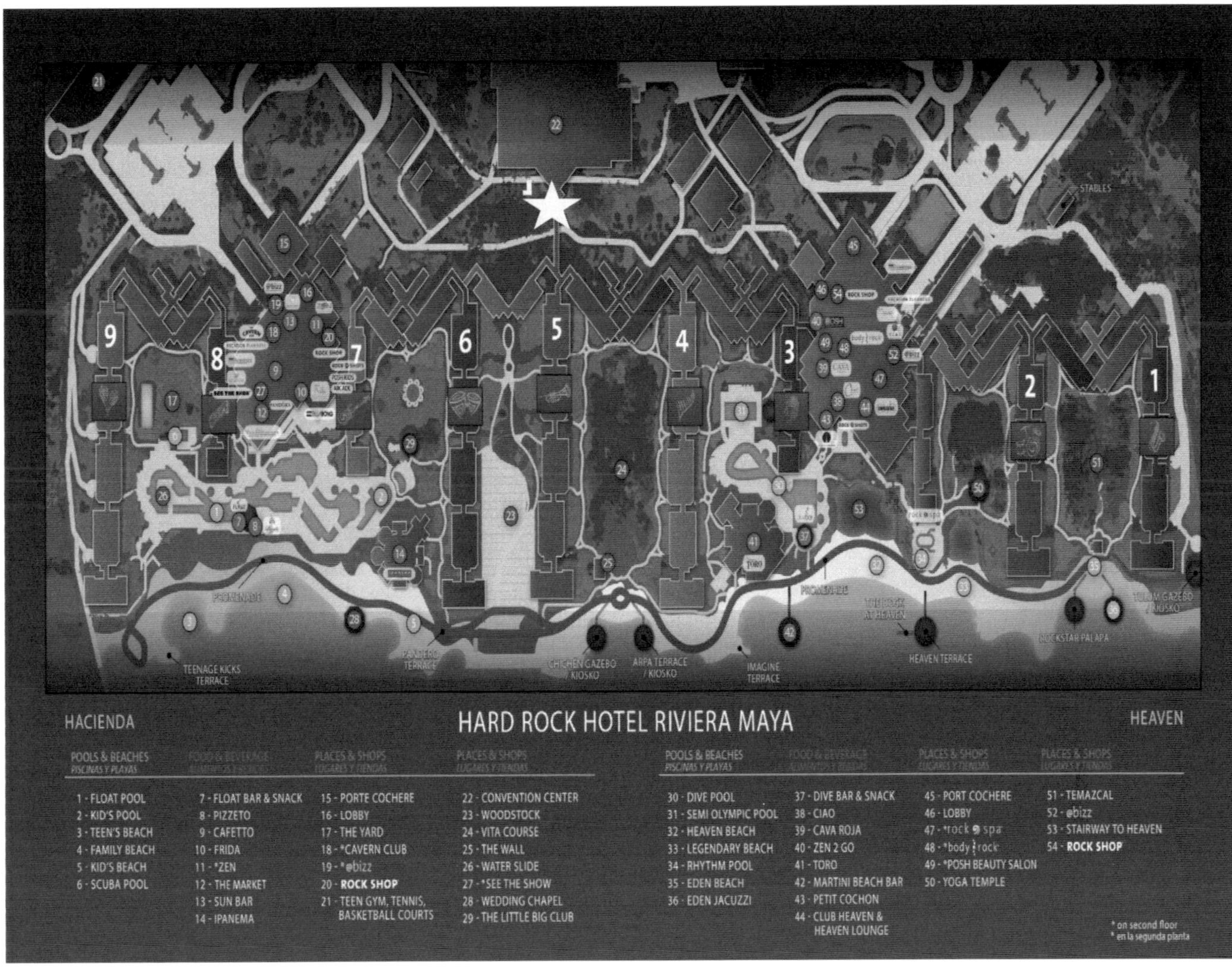

Convention Center

International Convention Center - Hard Rock Hotel Riviera Maya
Chetumal, Carr. Cancún - Tulum, Puerto Aventuras Km 72
Riviera Maya, Quintana.Roo.
ZIP Code 77710.

Under de patronage of

Coordinación General de Ganadería. SAGARPA

PLATINUM

CNG

Confederación Nacional de Organizaciones Ganaderas

GOLD

IDEXX LABORATORIES

CZV

CZ VETERINARIA

SILVER

ID.vet

Innovative Diagnostics

Biolife

STAND

QIAGEN

OTHER SPONSORS

Municipio San Juan del Río, Querétaro

GANADERIA RIVERA DE LAJAS

Ganadería Santa Cruz de Xarama

International Association for Paratuberculosis
2015 Linden Drive
Madison, Wisconsin 53706-1102
USA

ISBN 978-1-5108-7302-5

CURRAN ASSOCIATES INC.
proceedings.com